Journal of Neural Transmission

Supplementum 34

H. Bönisch, K.-H. Graefe, S. Z. Langer, and E. Schömig (eds.)

Recent Advances in Neuropharmacology

Springer-Verlag   Wien New York

Prof. Dr. H. Bönisch
Institute of Pharmacology and Toxicology, University of Bonn, Federal Republic of Germany

Prof. Dr. K.-H. Graefe
Institute of Pharmacology and Toxicology, University of Würzburg, Federal Republic of Germany

Dr. S. Z. Langer
Department of Biology, Synthélabo Recherche (LERS), Paris, France

Dr. E. Schömig
Institute of Pharmacology and Toxicology, University of Würzburg, Federal Republic of Germany

With 44 Figures

ISSN 0303-6995
ISBN-13:978-3-211-82300-2   e-ISBN-13:978-3-7091-9175-0
DOI: 10.1007/978-3-7091-9175-0

Preface

This volume is dedicated to Professor Ullrich Trendelenburg. It contains the proceedings of a symposium which was held in his honour on the occasion of his retirement and took place March 22–24, 1991 in Würzburg. Ullrich Trendelenburg was the head of the Department of Pharmacology at Würzburg University from 1968 till the end of March 1991. He is famous internationally for his contributions to the physiology and pharmacology of the autonomic nervous system, and his impact on pharmacology in general throughout the world has been outstanding. The various phases of his life and his career have been delineated recently by Youdim and Riederer (Journal of Neural Transmission; Suppl. 32, 1990).

The articles included in this volume reflect a considerable range of current research work dealing with various aspects of neuropharmacology, i.e., the field of research Ullrich Trendelenburg has influenced most. One or more authors of each chapter are either former or present students and coworkers or close friends of Ullrich Trendelenburg. The *first section* is devoted to the synthesis and metabolism of catecholamines as well as to the mechanisms by which amine transmitters are removed from the extracellular fluid; three chapters deal with the two types of extraneuronal uptake of catecholamines. The *second section* concentrates on the release of catecholamines in the peripheral and the central nervous system, the regulation of transmitter release and the noradrenaline-ATP co-transmission. The *third section* deals with the pharmacology of various receptors, including adrenoceptors, adenosine, 5-HT and glutamate receptors.

We thank the editors of the Journal of Neural Transmission for giving us the opportunity to publish the proceedings of this Symposium.

H. Bönisch
K.-H. Graefe
S. Z. Langer
E. Schömig

Würzburg, November 1991

Contents

Receptors and post-receptor events

Listed in Current Contents

Synthesis and inactivation of neurotransmitters

J Neural Transm (1991) [Suppl] 34: 3–9

The effects of electroconvulsive shock on catecholamine function in the locus ceruleus and hippocampus

N. Weiner, M. A. Hossain, and **J. M. Masserano**

Department of Pharmacology, C-236, University of Colorado Health Sciences Center,
Denver, CO, U. S. A.

Summary. Repeated electroconvulsive shock (ECS) treatment (once per day for 7 days) produced a significant increase in tyrosine hydroxylase activity, GTP-cyclohydrolase activity and tetrahydrobiopterin (BH_4) levels in the locus ceruleus and hippocampus from 1 to 4 days after the last treatment. These changes may be responsible for, or contribute to, the antidepressant effect of ECS treatment.

Introduction

Electroconvulsive shock (ECS) is the most effective therapy available for the treatment of endogenous depression. The antidepressant effects of ECS appear only after repeated treatments and persist for an extended period of time after the last treatment. The precise neurochemical mechanism underlying the antidepressant effects of ECS is not completely understood. A variety of biochemical changes are known to occur in noradrenergic neurones in the central nervous system after repeated ECS. These include an increase in norepinephrine turnover (Kety et al., 1967; Modigh, 1976), an increase in norepinephrine tissue concentrations (Kety et al., 1967), a decreased affinity for the high affinity uptake of norepinephrine (Hendley, 1976; Hendley and Welch, 1974), a decrease in the density of postsynaptic beta receptors (Bergstrom and Kellar, 1979), a decrease in norepinephrine-stimulated cyclic AMP accumulation (Vetulani and Sulser, 1975; Vetulani et al., 1976), an increase in alpha-1 adrenoceptor binding (Vetulani et al., 1983; Stockmeier et al., 1987), a decrease in alpha-2 adrenoceptor binding (Heal et al., 1981; Pilc and Vetulani, 1982), and an increase in norepinephrine and dopamine release (Glue et al., 1990). These changes in noradrenergic neurones are thought to be related to the antidepressant effects of ECS since they occur, for the most part, only after repeated ECS administration and do not occur after a single ECS treatment (Gleiter and Nutt, 1989).

Our laboratory has previously reported an increase in tyrosine hydroxylase activity in the central nervous system and adrenal glands of rats following repeated ECS (Masserano et al., 1981). The increases in tyrosine hydroxylase were initially apparent in the noradrenergic cell body regions of the central nervous system one day after the last ECS (1/day for 7 days) and progressed to the nerve terminal regions over a period of 1 to 8 days. We have extended these studies by evaluating the effects of repeated ECS (1/day for 7 days) on the levels of tetrahydrobiopterin (BH_4) and on the activity of the enzyme GTP-cyclohydrolase. BH_4 and GTP-cyclohydrolase were evaluated in the locus ceruleus and hippocampus of rats after ECS treatment. BH_4 is the cofactor for tyrosine hydroxylase and GTP-cyclohydrolase is the rate limiting enzyme in the biosynthetic pathway of BH_4.

Materials and methods

ECS administration

ECS was administered to male Sprague-Dawley rats by the application of a 300 mA current transorbitally for a duration of 0.2 seconds. ECS was administered either one time or once a day for a period of seven days. Five minutes, one day, four days, or eight days after the last ECS rats were anesthetized with halothane, decapitated and the locus ceruleus and hippocampus dissected out and frozen on dry ice. These brain areas were stored at $-70°C$ until assayed for tyrosine hydroxylase activity, GTP-cyclohydrolase activity or BH_4 levels.

Tyrosine hydroxylase assay

Tyrosine hydroxylase was determined by the coupled decarboxylase assay as modified in our laboratory (Masserano et al., 1981).

BH_4 assay

BH_4 was measured by HPLC analysis using fluorescence detection as described by Fukushima and Nixon (1980) and Lee and Mandell (1985).

GTP-cyclohydrolase assay

The activity of GTP-cyclohydrolase was assayed according to the method of Duch et al. (1984). Tissues were sonicated in 0.1 M Tris-HCl buffer (pH 7.8), containing 0.3 M KCl, 2.5 mM EDTA and 10% glycerol and centrifuged at $26,000 \times g$ for 20 min. The supernatant was filtered on a Sephadex G-25 column (0.5×6.5 cm) which had been equilibrated with the same buffer. The columns were washed initially with 0.3 ml of the buffer and the second wash (0.7 ml) was collected for assay of GTP-cyclohydrolase. The assay reaction mixture contained 0.2 ml of the eluate and 0.05 ml of 10 mM GTP.

The reaction was performed for 90 min at 37°C in the dark and terminated by the addition of 0.025 ml of a mixture of 1% I_2, 2% KI in 1 N HCl. The excess I_2 was removed by the addition of 2% ascorbic acid. The neopterin triphosphate formed by the enzymatic reaction was dephosphorylated by the addition of 0.025 ml of 1N NaOH followed by 1.5 units of bacterial alkaline phosphatase and incubated for an additional 1 h at 37°C in the dark. The reaction was terminated by the addition of 0.050 ml 1 N acetic acid. Following centrifugation, neopterin was measured by HPLC analysis as described by Lee and Mandell (1985).

Results

The effects of ECS on tyrosine hydroxylase activity in the locus ceruleus and hippocampus at 1, 4 and 8 days following repeated ECS treatment are shown in Fig. 1. Tyrosine hydroxylase activity was significantly increased in the locus ceruleus 1 and 4 days following the last treatment and had returned to control levels by 8 days. The increase in tyrosine hydroxylase activity in the hippocampus was delayed until 4 days and returned to control levels by 8 days.

The effects of ECS on GTP-cyclohydrolase activity in the locus ceruleus and hippocampus at 1, 4, and 8 days following repeated ECS treatment are shown in Fig. 2. GTP-cyclohydrolase activity was significantly increased in the locus ceruleus 1 day following the last treatment and had returned to

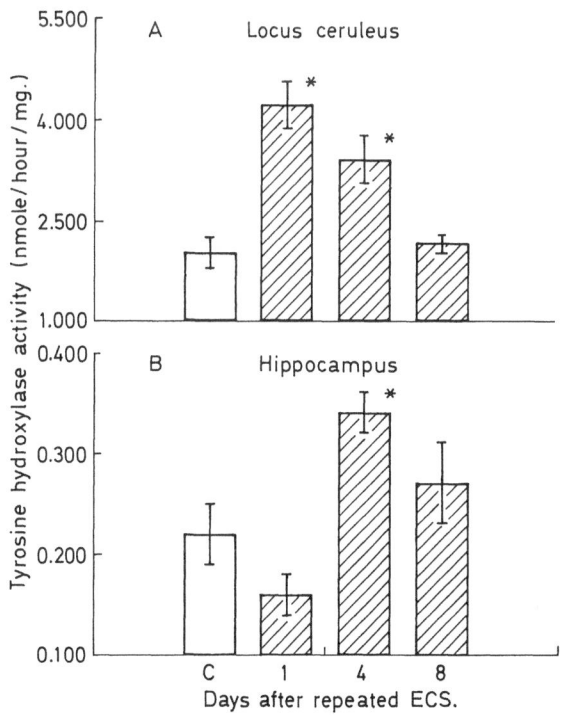

Fig. 1. Tyrosine hydroxylase activity in the locus ceruleus and hippocampus at 1, 4, and 8 days after repeated ECS treatment (1/day for 7 days). Significantly different from control; *P < 0.05

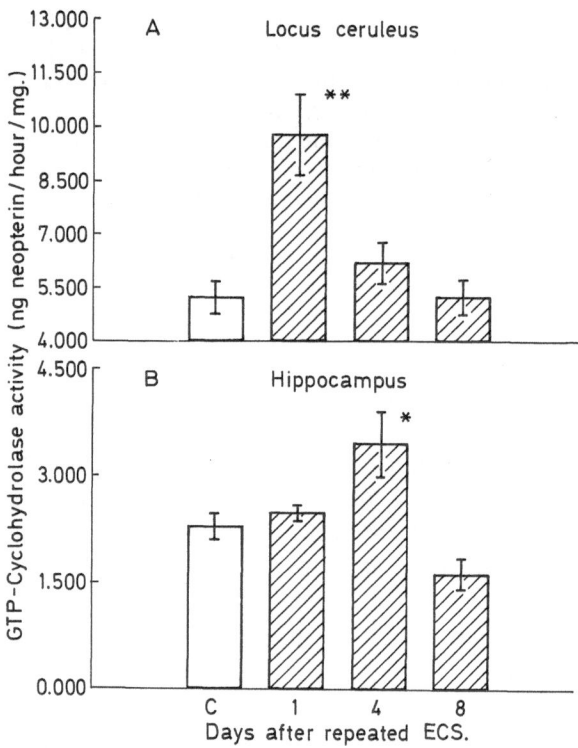

Fig. 2. GTP-cyclohydrolase activity in the locus ceruleus and hippocampus at 1, 4, and 8 days after repeated ECS treatment (1/day for 7 days). Significantly different from control; *P < 0.05, **P < 0.01

control levels by 4 days. The increase in GTP-cyclohydrolase in the hippocampus was delayed until 4 days and returned to control levels by 8 days.

The effects of ECS on BH_4 levels in the locus ceruleus and hippocampus 5 min after a single ECS and at 1 day after repeated ECS treatment are shown in Fig. 3. BH_4 levels were significantly increased in the locus ceruleus and hippocampus 1 day after repeated ECS treatment with no change occurring 5 min after a single ECS.

Discussion

One day and four days after the last ECS treatment, tyrosine hydroxylase activity was increased in the locus ceruleus, a noradrenergic cell body region of the brain. Tyrosine hydroxylase activity was significantly increased in the hippocampus, a noradrenergic nerve terminal region of the brain, at four days after the last ECS treatment. The noradrenergic cell bodies of the locus ceruleus send projections to the hippocampus. Based on the rate of tyrosine hydroxylase transport in the noradrenergic neurones of the central nervous system (approximately 2 to 7 mm per day) (Zigmond, 1978; Black,

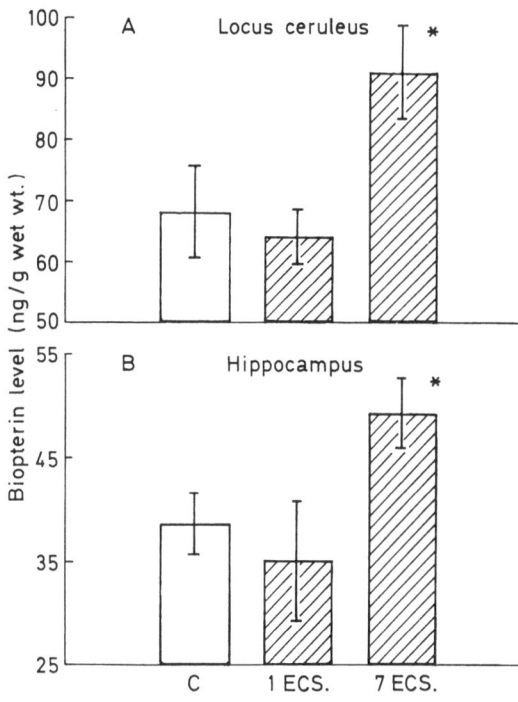

Fig. 3. BH$_4$ levels in the locus ceruleus and hippocampus 5 minutes after a single ECS and 1 day after repeated ECS treatment (1/day for 7 days). Significantly different from control; *P < 0.05

1975), it appears that the increase in tyrosine hydroxylase enzyme protein produced in the locus ceruleus after repeated ECS is transported to the hippocampus.

Similar results were obtained in the activity of GTP-cyclohydrolase after repeated ECS. GTP-cyclohydrolase is the rate-limiting enzyme in the pathway concerned with the synthesis of tetrahydrobiopterin, the cofactor for the enzyme tyrosine hydroxylase. We report that the repeated administration of ECS produces an increase in GTP-cyclohydrolase activity in the locus ceruleus one day after the last treatment and in the hippocampus four days after the last treatment. These changes in GTP-cyclohydrolase parallel the changes in the activity of tyrosine hydroxylase in these two brain regions after ECS. The activity of both tyrosine hydroxylase and GTP-cyclohydrolase return to baseline levels within eight days. Other laboratories report that GTP-cyclohydrolase can be regulated in the adrenal gland and brain by various treatments. Viveros et al. (1981) reported that following adrenal stimulation, GTP-cyclohydrolase activity increased in the adrenal gland and was significantly reduced after denervation. Levine et al. (1981) found that the lesion of the substantia nigra by 6-hydroxydopamine treatment produced in the striatum a 93% reduction of tyrosine hydroxylase, a 67% reduction of GTP-cyclohydrolase and a 73% reduction of BH$_4$. Since most of the BH_4 and GTP-cyclohydrolase in the nigrostriatal tract is located in dopaminergic neurons these data suggest that there is a up-regulation of BH_4 levels and GTP-cyclohydrolase activity in response to the severe depletion of tyrosine hydroxylase.

We also found that the levels of BH_4 were increased in the locus ceruleus and hippocampus after repeated ECS. BH_4 was increased 34% in the locus ceruleus and 28% in the hippocampus 1 day after the last ECS. The time course of this increase in BH_4 levels was different in the hippocampus than that which occurred with the enzymes. BH_4 levels in the hippocampus were elevated at one day following the last ECS treatment at a time when the activity of the enzymes GTP-cyclohydrolase and tyrosine hydroxylase were still at control levels. It is possible that this increase in BH_4 levels may have persisted for an even greater period of time. Few changes have been reported to occur in BH_4 levels in the brain after various manipulations. Mandell and coworkers (1980) examined the effects of twenty-seven different drug treatments on BH_4 levels in the striatum. The major reproducible change produced was a 25 to 30% decrease in BH_4 levels after treatment with d-amphetamine.

The increases in tyrosine hydroxylase activity, GTP-cyclohydrolase activity and BH_4 levels that are apparent for at least 1 to 4 days after repeated ECS may be responsible for, or contribute to, the antidepressant effect of this treatment. The possibility that other brain areas and the adrenal medulla may also undergo similar changes in GTP-cyclohydrolase activity and BH_4 levels after repeated ECS is currently under investigation.

References

Bergstrom DA, Kellar KJ (1979) Effect of electroconvulsive shock on monoaminergic receptor binding sites in rat brain. Nature 278: 464–466

Black IB (1975) Increased tyrosine hydroxylase activity in frontal cortex and cerebellum after reserpine. Brain Res 95: 170–176

Duch DS, Bowers SW, Woolf JH, Nichol CA (1984) Bioterin cofactor biosynthesis: GTP cyclohydrolase, neopterin and biopterin in tissues and body fluids of mammalian species. Life Sci 35: 1895–1901

Fukushima T, Nixon JC (1980) Analysis of reduced forms of biopterin in biological tissues and fluids. Anal Biochem 102: 176–188

Gleiter CH, Nutt DJ (1989) Chronic electroconvulsive shock and neurotransmitter receptors — an update. Life Sci 44: 985–1006

Glue P, Costello MJ, Pert A, Mele A, Nutt DJ (1990) Regional neurotransmitter responses after acute and chronic electroconvulsive shock. Psychopharmacology 100: 60–65

Heal DJ, Akagi H, Bowdler JM, Green AR (1981) Repeated electroconvulsive shock attenuates clonidine-induced hypoactivity in rodents. Eur J Pharmacol 75: 231–237

Hendley ED (1976) Electroconvulsive shock and norepinephrine uptake kinetics in the rat brain. Psychopharmacol Commun 2(1): 17–25

Hendley ED, Welch BL (1974) Electroconvulsive shock: sustained decrease in norepinephrine uptake affinity in a reserpine model of depression. Life Sci 16: 45–54

Kety SS, Javoy F, Thierry A, Julou L, Glowinski J (1967) A sustained effect of electroconvulsive shock on the turnover of norepinephrine in the central nervous system of the rat. Proc Natl Acad Sci 58: 1249–1254

Lee EHY, Mandell AJ (1985) Relationships between drug-induced changes in tetahydrobiopterin and biogenic amine concentrations in rat brain. J Pharmacol Exp Ther 234: 141–146

Levine RA, Miller LP, Lovenberg W (1981) Tetrahydrobiopterin in straitum: localization in dopamine nerve terminals and role in catecholamine synthesis. Science 214: 919–921

Mandell AJ, Bullard WP, Yellin JB, Russo PV (1980) The influence of D-amphetamine of rat brain striatal reduced biopterin concentration. J Pharmacol Exp Ther 213: 569–574

Masserano JM, Takimoto GS, Weiner N (1981) Electroconvulsive shock increases tyrosine hydroxylase activity in the brain and adrenal gland of the rat. Science 214: 662–665

Modigh K (1976) Long-term effects of electroconvulsive shock therapy on synthesis turnover and uptake of brain monoamines. Psychopharmacology 49: 179–185

Pilc A, Vetulani J (1982) Depression by chronic electroconvulsive treatment of clonidine hypotherma and [^{3}H]clonidine binding to rat cortical membranes. Eur J Pharmacol 80: 109–113

Stockmeier CA, McLeskey SW, Blendy JA, Armstrong NR, Kellar KJ (1987) Electroconvulsive shock but not antidepressant drugs increases α_1-adrenoceptor binding sites in rat brain. Eur J Pharmacol 139: 159–166

Vetulani J, Sulser F (1975) Action of various antidepressant treatments reduces reactivity of noradrenergic cyclic AMP-generating system in limbic forebrain. Nature 257: 495–496

Vetulani J, Stawarz RJ, Dingell JV, Sulser F (1976) A possible common mechanism of action of antidepressant treatments. Naunyn-Schmiedebergs Arch Pharmacol 293: 109–114

Vetulani J, Antkiewicz-Michaluk L, Rokosz-Pelc A, Pilc A (1983) Chronic electroconvulsive treatment enhances the density of [^{3}H]prazosin binding sites in the central nervous system of the rat. Brain Res 275: 392–395

Viveros OH, Lee CL, Abou-Donia MM, Nixon JC, Nichol CA (1981) Biopterin cofactor biosynthesis: independent regulation of GTP cyclohydroxylase in adrenal medulla and cortex. Science 213: 349–350

Zigmond RE (1978) Tyrosine hydroxylase activity in noradrenergic neurons of the locus coeruleus after reserpine administration: sequential increase in cell bodies and nerve-terminals. J Neurochem 32: 23–29

Authors' address: N. Weiner, MD, University of Colorado Health Sciences Center, Department of Pharmacology, C-236, 4200 East 9th Ave, Denver, CO 80262, U.S.A.

J Neural Transm (1991) [Suppl] 34: 11–17

Molecular aspects of the neuronal noradrenaline transporter

H. Bönisch, G. Paulus, G. Martiny-Baron, B. Lingen, and **D. Coppeneur**

Department of Pharmacology and Toxicology, University of Bonn,
Federal Republic of Germany

Summary. The neuronal noradrenaline transporter was partially purified by means of low and high pressure liquid chromatography using anion exchange, gel filtration and lectin affinity columns. A protein characterized by a molecular weight of 50–53 kilodalton was enriched; it may represent the transporter or a component of it. In addition, a RNA fraction characterized by a mean size of 2 kilobases was isolated from PC12 rat phaeochromocytoma cells and from bovine adrenal medulla; this RNA fraction caused expression of the noradrenaline transporter after microinjection into Xenopus laevis oocytes.

Introduction

Membrane transport is one important way by which neurotransmitters are inactivated after their release into the synaptic gap. At least two transport processes are known to participate in the inactivation of noradrenaline, "uptake$_1$" located on the plasma membrane of noradrenergic neurones, and "uptake$_2$" located on the effector cells; both systems were first described by Iversen (see e.g. Iversen, 1967).

Transport of noradrenaline via "uptake$_1$" is absolutely dependent on Na^+ and Cl^-, competitively inhibited by cocaine, nisoxetine and the tricyclic antidepressant desipramine and irreversibly blocked by the haloalkylamine xylamine (for review see Graefe and Bönisch, 1988). The recognition site for noradrenaline and desipramine has been studied by means of the binding of tritiated desipramine (^{3}H-DMI) to intact and solubilized membranes of PC12 rat phaeochromocytoma cells (which possess "uptake$_1$") and bovine adreno-medullary membranes (Bönisch and Harder, 1986; Schömig and Bönisch, 1986; Bönisch and Michael-Hepp, 1989).

Recently the noradrenaline transporter has been partially purified (Bönisch et al., 1990); in this study it has also been shown that a tritiated xylamine derivative (^{3}H-DMX) labelled two proteins (a 32 and a 53 kDa protein) which obviously were fragments of the transporter. A part of the present study is a continuation of our attempts to further purify the transporter. Another aim of the present study was to enrich a RNA fraction which encodes the transporter protein and to express the noradrenaline transport system in Xenopus laevis oocytes (these results have been reported in part elsewhere; Coppeneur et al., 1991).

Materials and methods

Purification of the noradrenaline transporter by column chromatography

Plasma membranes from bovine adrenal medulla were isolated as described by Bönisch et al. (1990). The isolated membranes were solubilized using the non-ionic detergent digitonin (1%) at a detergent/protein ratio of 10 (for further details see Schömig and Bönisch, 1986). The solubilized membranes were diluted with buffer A (20 mM Tris-HCl pH 7.4, 1 mM dithiothreitol, 1 mM EDTA, 1 mM phenylmethylsulfonyl fluoride, 0.1% digitonin) to a digitonin concentration of 0.5% and then applied to a low pressure anion-exchange column (DEAE-Sepharose FF preequilibrated with buffer A). Elution was carried out with 100 mM NaCl (in buffer A). The noradrenaline transporter was detected by binding of tritiated desipramine (see below). The transporter-containing fractions were pooled, diluted to 15 mM NaCl and applied to FPLC anion-exchange column (Mono Q HR 5/5 preequilibrated with buffer A containing 15 mM NaCl). Elution was carried out with a gradient of 15–300 mM NaCl in buffer A. The fractions containing the noradrenaline transporter were pooled, diluted to 15 mM NaCl, again applied to the preequilibrated Mono Q HR column and eluted with 300 mM NaCl in buffer A. This concentrated fraction was then mixed with a lectin affinity resin (WGA-Sepharose 6 MB) and buffer A containing 1 M NaCl. After washing a noradrenaline transporter-containing fraction was eluted from the resin with buffer A containing 100 mM N-acetylglucosamine (besides 1 M NaCl). In some experiments an additional purification step was introduced before the lectin affinity resin; i.e., the concentrated fraction obtained from the Mono Q HR column was applied to a Superose 6 HR 10/30 gel filtration column (preequilabrated with 150 mM NaCl in buffer A containing 0.02% sodium azide) and the transporter was eluted with this preequlibration buffer using a flow rate of 0.2 ml/min.

Binding of ^{3}H-desipramine (^{3}H-DMI) to membranes, solubilized membranes or to fractions obtained during the purification were performed by a filtration assay using 1 nM ^{3}H-DMI and buffer A containing 1 M NaCl. Nisoxetine (10 µM) was used to determine the non-specific binding of ^{3}H-DMI (for further details, see Bönisch and Harder, 1986).

Expression of the transporter in Xenopus laevis oocytes

Cultured PC12 cells (Harder and Bönisch, 1984) and freshly isolated bovine adrenal medullae were dissolved in guanidinium thiocyanate (4 M) containing lauroyl sarcosine (2%). From this solution the RNA was isolated (Coppeneur et al., 1991). The RNA was centrifuged through a CsCl layer (5.7 M) at 150,000 × g for 22 h, precipitated witch 70% ethanol and size fractioned using a 10–30% sucrose gradient centrifugation (150,000 × g, 16 h). Poly(A$^+$)mRNA was separated from total RNA by oligo(dT) chromatography (Maniatis et al., 1982).

Uptake of ^{3}H-noradrenaline by oocytes

Oocytes were isolated with collagenase (2 mg/ml) from ovarian lobes of Xenopus laevis and maintained in culture using modified Barth's solution (MBS) as described by Colman (1984). The MBS consisted of (in mM): 88 NaCl, 1 KCl, 0.8 MgSO$_4$, 0.33 Ca(NO$_3$)$_2$, 0.4 CaCl$_2$, 2.4 NaHCO$_3$, 10 Hepes/NaOH pH 7.5. Healthy oocytes were injected with 60 nl RNA (1 mg/ml) and mainatined in culture for about 3 days (in MBS containing penicillin 10,000 U/ml and streptomycin 10 µg/ml). To measure the uptake of ^{3}H-noradrenaline (^{3}H-NA), oocytes were incubated (for 2 h at 22°C) in MBS containing 10 µM pargyline (to inhibit monoamine oxidase) and 10 µM U-0521 (to inhibit catechol-O-methyl transferase), 1 mM ascorbic acid and 100 nM ^{3}H-NA (specific activity 14.2 Ci/mmol). The uptake of ^{3}H-NA which was sensitive to nisoxetine (10 µM) was regarded as specific. Uptake experiments were terminated by washing the oocytes with ice-cold MBS, dissolving the oocytes in 1% Triton-X-100 and determining their tritium content.

Materials

^{3}H-(-)Noradrenaline (NEN, Dreieich, FRG), U-0521 (Upjohn, Kalamazoo, MI, USA), nisoxetine (Lilly, Indianapolis, IN, USA), iprindol (Wyeth-Pharma, Münster, FRG), oxaprotiline (Ciba-Geigy, Basel, Switzerland); all other compounds were either from Sigma (München, FRG) or from Merck (Darmstadt, FRG). Columns and resins were from Pharmacia (Freiburg, FRG).

Results

Purification of the noradrenaline transporter

Binding of tritiated desipramine (^{3}H-DMI) to the tricyclic antidepressant binding site (which is identical with the substrate recognition site) of the neuronal noradrenaline transporter was used to follow the purification of this membrane protein. Among several detergents tested, the non-ionic detergent digitonin proved to be the best. Solubilization of bovine adreno-medullary plasma membranes with digitonin resulted in an 1.4-fold increase in the specific binding of ^{3}H-DMI (Table 1), i.e., in a partial enrichment of the ^{3}H-DMI binding site.

To further purify the transporter protein, several chromatographic procedures were investigated. No purification of the digitonin solibilized bovine noradrenaline transporter was obtained with an affinity resin consisting of an imipramine derivative (methylamino imipramine) coupled (via a spacer arm) to an agarose matrix, although this affinity column had been used successfully to partially purify the noradrenaline transporter from rat phaeochromocytoma (PC12) membranes (Bönisch et al., 1990). Therefore, conventional purification procedures were used. Purification was started with a combination of two anion exchange resins, the low pressure

DEAE-Sepharose FF and the FPLC resin Mono Q HR 5/5. The transporter was eluted from the DEAE-Sepharose FF with a single step gradient of 100 mM NaCl (see Methods); the eluate contained an 11-fold purified transporter protein (Table 1). When this prepurified fraction (diluted to 15 mM NaCl) was loaded on a Mono Q HR 5/5 column and the column was eluted by means of a linear NaCl gradient (15–300 mM), the eluate fraction obtained with 90–150 mM NaCl contained the transporter protein with a mean ^{3}H-DMI binding activity of about 45 pmol/mg (Table 1); i.e., using this protocol, a 23-fold purification of the transporter (compared to intact membranes) was achieved (Table 1). The solubilized transporter was further purified by lectin affinity chromatography with WGA-Sepharose 6 MB, resulting in about 105-fold purification; the final protein recovery was about 0.1% of the starting material (Table 1). When the second anion exchange column (Mono Q HR 5/5) was omitted, the final purification after the WGA-sepharose column was only 90.5-fold, and the protein recovery under these conditions was about 0.2%. When the transporter protein within the eluate from the second anion exchange column was further purified by means of a gel filtration procedure using Superose 6 HR as support, the specific ^{3}H-DMI binding activity within the active fraction increased from 44.5 pmol/mg (Table 1) to 105 pmol/mg. When this fraction was applied to the WGA-sepharose resin, no further ^{3}H-DMI binding activity was detectable in the N-acetylglucosamine eluate of the column. However, analysis of the proteins of the eluate fraction by SDS-polyacrylamid gelelectrophoresis showed an enrichment of a protein characterized by a molecular size of about 50–54 kDa. Only four additional proteins with molecular weights of about 100, 75, 26 and 15 kDa were detected on the gel.

Enrichment of the transporter-RNA and expression of the transporter in oocytes

RNA isolated from PC12 cells and from bovine adrenal medulla was size fractionated and the fractions were microinjected into Xenopus laevis oocytes. Three days after injection, measurement of nisoxetine-sensitive uptake of ^{3}H-noradrenaline (^{3}H-NA) into the injected oocytes showed that one RNA fraction from either PC12 cells or bovine adrenal medulla, characterized by a size of about 2 kilobases, caused significant ^{3}H-NA uptake (1,86 ± 0.24 (n = 18) and 1,0 ± 0.25 (n = 24) fmol/h per oocyte, respectively). Interestingly, with size selected RNA from PC12 cells (but not from bovine adrenal medulla) a sigificant uptake of ^{3}H-NA was also obtained with a 4 kb RNA fraction. No specific uptake of ^{3}H-NA was obtained in non-injected or water-injected oocytes, indicating a lack of endogenous noradrenaline transport activity.

As shown in Fig. 1, the uptake of ^{3}H-NA induced by the 2 kb RNA from bovine adrenal medulla remained unaltered in the presence of iprindol

Table 1. Partial purification of the NA-transporter (^{3}H-DMI binding site) of the plasma membrane of bovine adrenal medulla

	^{3}H-DMI binding (pmol/mg)	Purification factor	Protein recovery (%)
Plasma membranes	1.9	1	100
Solubilized membranes (digitonin 0.5%)	2.7	1.4	49
Chromatographic steps:			
a) anion exchange (DEAE-sepharore FF)	21.7	11.4	6.4
b) ″ ″ (Mono Q HR 5/5; FPLC)	44.5	23.4	1.2
c) lectin affinity (WGA-sepharose 6MB)	200	105	0.1

Shown are means of two purification experiments

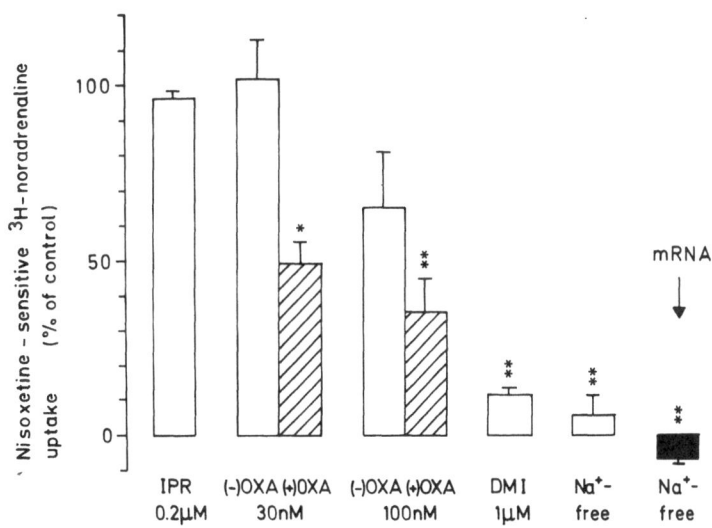

Fig. 1. RNA-induced expression of the noradrenaline transport system in oocytes. Xenopus laevis oocytes were microinjected with a 2 kb RNA fraction (from bovine adrenal medulla) or a polyadenylated mRNA isolated from this fraction. After three days, uptake of ^{3}H-NA (100 nmol/l) was determined as described in Methods. Shown are mean values ± SEM of 7–29 oocytes per group. *P < 0.05, **P < 0.001 compared with controls. *IPR* iprindol, *OXA* oxaprotiline, *DMI* desipramine

(0.2 μM), but was markedly inhibited by desipramine (1 μM) and stereoselectively reduced by the (+)-enantiomer of oxaprotiline (Fig. 1). The ^{3}H-NA uptake induced by the 2 kb RNA or by a 2 kb poly(A$^+$)mRNA (isolated from the RNA) was dependent on Na$^+$, since no uptake was observed when Na$^+$ in the incubation buffer was replaced by Tris$^+$ (Fig. 1).

Discussion

By means of low and high pressure liquid chromatography using anion exchange chromatography followed by gel filtration and affinity chromatography to a lectin support (WGA-Sepharose) it was possible to specifically enrich a protein characterized by a molecular size of about 50–54 kDa. The binding to the lectin column indicates that this protein is a glycoprotein. Moreover, the results suggest that the protein probably represents the neuronal noradrenaline transporter or at least a fragment of it. This suggestion is supported by the fact that a membrane protein (from PC12 cells and from bovine adrenal medulla) of this size was selectively labelled by tritiated desmethylxylamine (^{3}H-DMX), a halo-alkylamine which covalently binds to the neuronal noradrenaline transporter (Bönisch and Michael-Hepp, 1990; Bönisch et al., 1990). In addition, Howard et al. (1990) recently described a 54 kDa PC12 membrane protein which was covalently labelled by ^{3}H-xylamine to represent the noradrenaline transporter. We can identify a ^{3}H-DMX-labelled 53 kDa protein as a single spot (characterized by an isoelectric point of about 6.5) in a 2-dimensional gel (Martiny-Baron and Bönisch, unpublished). Thus, using this technique and protein fractions containing the partially purified transporter, it should be possible to isolate enough of this 53 kDa membrane protein for microsequencing and to deduce from such a sequence oligonucleotides for cDNA cloning of the transporter.

The present data indicate that we have isolated a 2 kb RNA fraction from both, PC12 cells and bovine adrenal medulla, which contains the message for the synthesis of the noradrenaline transporter. Microinjection into oocytes of the 2 kb RNA fraction induced specific ^{3}H-noradrenaline uptake by the neuronal noradrenaline transporter, since it was sensitive to nisoxetine, desipramine, (+)oxaprotiline and to sodium deprivation. The transport was not due to an induction of an endogenous adrenaline transport, since the uptake of ^{3}H-noradrenaline was not sensitive to iprindol, a selective inhibitor of the adrenaline transporter (Pimoule et al., 1987).

Using the 2 kb RNA fraction a cDNA library can be constructed und the cDNA encoding the noradrenaline transporter can be isolated e.g. by means of expression cloning using xenopus laevis oocytes as recently accomplished for the Na$^+$/glucose cotransporter (Hediger et al., 1987).

Note added

When this manuscript had been written, the expression cloning of the human noradrenaline transporter was published [Pacholczyk T, Blakely RD, Amara SG (1991) Nature 350:350–354]. The DNA sequence of the cloned transporter is smaller than 2 kb and the sequence predicts a glycoprotein of about 69 kDa.

Acknowledgement

This work was supported by the Deutsche Forschungsgemeinschaft.

References

Bönisch H, Harder R (1986) Binding of ^{3}H-desipramine to the neuronal noradrenaline carrier of rat phaeochromocytoma cells (PC12 cells). Naunyn-Schmiedebergs Arch Pharmacol 334: 403–411

Bönisch H, Michael-Hepp J (1989) Binding of ^{3}H-desipramine to the noradrenaline carrier of the plasma membrane of bovine adrenal medulla. Naunyn-Schmiedebergs Arch Pharmacol 340 [Suppl]: R38

Bönisch H. Michael-Hepp J (1990) Labelling of the neuronal noradrenaline-carrier with a xylamine derivative. Naunyn-Schmiedebergs Arch Pharmacol 341 [Suppl]: R82

Bönisch H, Martiny-Baron G, Blum B, Michael-Hepp J (1990) Biochemical characterization and purification of the neuronal sodium-dependent noradrenaline transporter. J Neural Transm [Suppl] 32: 413–419

Colman A (1984) Translation of eukaryotic messenger RNA in Xenopus oocytes. In: Hames BD, Higgins SJK (eds) Transcription and translation. A practical approach. IRL Press, Oxford, pp 271–302

Coppeneur D, Lingen B, Sanders G, Dabauvalle M-C, Bönisch H (1991) Expression of the neuronal noradrenaline transporter. Naunyn-Schmiedebergs Arch Pharmacol 343: 334–336

Graefe K-H, Bönisch H (1988) The transport of amines across the axonal membranes of noradrenergic and dopaminergic neurones. In: Trendelenburg U, Weiner N (eds) Catecholamines I. Springer, Berlin Heidelberg New York Tokyo, pp 193–245

Harder R, Bönisch H (1984) Large-scale preparation of plasma membrane vesicles from PC-12 pheochromocytoma cells and their use in noradrenaline transport studies. Biochim Biophys Acta 775: 95–104

Hediger MA, Coady MJ, Ikeda TS, Wright EM (1987) Expression cloning and cDNA sequencing of the Na^+/glucose co-transporter. Nature 330: 379–381

Howard BD, Cho AK, Zhang M-B, Koide M, Lin S (1990) Covalent labeling of the cocaine-sensitive catecholamine transporter. J Neurosci Res 26: 149–158

Iversen LL (1967) The uptake and storage of noradrenaline in sympathetic nerves. Cambridge University Press, Cambridge

Maniatis T, Fritsch EF, Sambropok J (1982) Molecular cloning: a laboratory manual. Cold Spring Harbor Laboratory, Cold Spring Harbour, New York

Pimoule C, Schoemaker H, Langer SZ (1987) ^{3}H-Desipramine labels with high affinity the neuronal transporter for adrenaline in the frog heart. Eur J Pharmacol 137: 277–280

Schömig E, Bönisch H (1986) Solubilization and characterization of the ^{3}H-desipramine binding site of rat phaeochromocytoma cells (PC12 cells). Naunyn-Schmiedebergs Arch Pharmacol 37: 412–417

Authors' address: Dr. H. Bönisch, Department of Pharmacology and Toxicology, University of Bonn, Reuterstrasse 2b, D-W-5300 Bonn 1, Federal Republic of Germany

J Neural Transm (1991) [Suppl] 34: 19–25

Energy requirements for the basal efflux of noradrenaline and its metabolites from adrenergic varicosities

U. Trendelenburg, H. Russ, and **E. Schömig**

Department of Pharmacology and Toxicology, University of Würzburg, Federal Republic of Germany

Summary. The combination of hypoxia plus glucose deprivation or of hypoxia plus lactate induces carrier-mediated outward transport of ^{3}H-noradrenaline in the rat vas deferens. Lactate efflux is higher from atria than from vas deferens. Hence, the much lower contribution by outward transport to the spontaneous efflux of ^{3}H-noradrenaline in vas deferens than atria is likely to be due to a better supply of oxygen (and perhaps also glucose) to the ^{3}H-noradrenaline-storing varicosities in vas deferens than in atria.

Abbreviations

COMT catechol-O-methyl transferase, *DOMA* dihydroxymandelic acid, *DOPEG* dihydroxyphenylglycol, *FRL* fractional rate of loss, *MAO* monoamine oxidase

The composition of the spontaneous efflux of tritium from vasa deferentia and atria

In experiments with atria and vasa deferentia of the rat (loaded with 0.2 μmol/l ^{3}H-noradrenaline for 60 min and then washed out with amine-free solution for 115 min), we noticed that the spontaneous efflux of tritium was much higher in atria than in vasa deferentia (Schömig et al., 1989). Analysis of this phenomenon revealed that the difference was due to a corresponding difference for the spontaneous efflux of ^{3}H-noradrenaline (Table 1). Moreover, in these experiments we calculated the contribution to the outflow of ^{3}H-noradrenaline by (desipramine-resistant) outward *diffusion* and by carrier-mediated (desipramine-sensitive) outward *transport*. The latter was about four times higher in atria than in vasa deferentia, although outward diffusion was similar in both tissues (Schömig et al., 1989).

Table 1. The spontaneous efflux of tritium from rat vasa deferentia and atria

	Vasa deferentia		Atria	
n		24		19
FRL ^{3}H	0.00619	(0.00551; 0.00695)	0.00909	(0.00855; 0.00968)
FRL ^{3}H-NA	0.00199	(0.00185; 0.00215)	0.00512	(0.00472; 0.00551)
FRL ^{3}H-DOPEG	0.00377	(0.00312; 0.00455)	0.00333	(0.00306; 0.00364)

Geometric means (with 95% confidence limits in parentheses) of the FRL (= rate of efflux divided by tissue tritium content determined at the same time) for tritium, ^{3}H-noradrenaline and ^{3}H-DOPEG (in min^{-1}). Rats were pretreated with reserpine (to block vesicular uptake) and pargyline (to block MAO). COMT was inhibited by the presence of 100 µmol/l U-0521. The tissues were exposed to 0.2 µmol/l ^{3}H-noradrenaline for 60 min and then washed with amine-free solution. Spontaneous efflux was collected from 115 to 120 min. Results taken from Schömig et al. (1989)

The heterogeneity of the neuronal distribution of ^{3}H-noradrenaline in the incubated vas deferens of the rat

Both by autoradiography (Azevedo et al., 1990) and by analysis of the spontaneous efflux of ^{3}H-noradrenaline and its ^{3}H-metabolites (Schömig et al., 1990), it has been established that, during incubation with ^{3}H-noradrenaline, the amine is preferentially stored in those varicosities that are very close to the surface of the tissue. This very heterogeneous distribution of the exogenous amine is due to the very avid uptake$_1$ which is able to generate a steep steady-state concentration gradient for ^{3}H-noradrenaline in the extracellular space (Schömig et al., 1991).

The density of the adrenergic innervation is about seven times higher in the vas deferens than in atria of the rat (Hermann and Graefe, 1977). Hence, it is highly likely that the concentration gradient (see above) is much flatter in atria than in vasa deferentia. This must mean that the average depth of penetration of the ^{3}H-amine into the tissue may well be considerably greater in atria than in vasa deferentia. Hence, we face the possibility that, in rat atria, a high percentage of the ^{3}H-amine is stored in varicosities that potentially suffer from a poor supply of oxygen and/or glucose.

The energy requirements for the uptake and retention of substrates of uptake$_1$

Earlier studies date back to a time when it was not yet generally accepted that "initial rates" of uptake must be measured, if one is interested in uptake$_1$. Hence, in most of the studies summarized in Table 2 incubation with the substrate of uptake$_1$ was so long that the *outward* flux of the substrate must have affected the results. While Hamberger (1967) measured the restitution of histofluorescence for catecholamines, all others measured the "accumulation" of the ^{3}H-substrate in the tissue. For some details of experimental procedure, see Table 2; for further details (concentrations of,

Table 2. Inhibition of "neuronal uptake"[a] by impairment of the energy requirements. Lack of effect is indicated by "0", inhibition by " ↓ ". Absence of symbol means "not studied".

	Hamberger (1967)[b]	Wakade and Furchgott (1968)[c]	Paton (1968)[d]	Paton (1972)[e]
A. Prevention or inhibition of glycolysis				
Omission of glucose	0	0	0	0
+2-Deoxy-D-glucose				0
+IAA[f]		0	↓	0
B. Inhibition of oxidative phosphorylation				
Omission of oxygen $(-O_2)$	0	0	0	0
+DNP[g]	0	0	0	0
+cyanide	0			
C. Combinations of "B + A"				
$-O_2$ −glucose	↓	↓	↓	↓
+DNP −glucose	↓	↓	↓	↓
+cyanide −glucose	↓			
+DNP +IAA		↓	↓	
+DNP −glucose		↓		↓
$-O_2$ +IAA		↓		↓
$-O_2$ +2-deoxy-glucose				↓

[a] Most results reflect "accumulation" or "retention" rather than "unidirectional uptake"; see text.
[b] Rat vas deferens after pretreatment with reserpine; incubation with α-methyl-noradrenaline for 20–30 min; measured was the restitution of histofluorescence.
[c] Guinea-pig left atrium, MAO inhibited; incubation with ^{3}H−noradrenaline for 5 min.
[d] Rat uterine horn; incubation with ^{3}H-metaraminol for 45 min; pretreatment of animals with reserpine failed to affect results.
[e] Rat vas deferens, MAO intact; incubation with ^{3}H-noradrenaline for 30 min.
[f] iodoacetic acid
[g] dinitrophenol

and length of pre-incubation with, the inhibitors), the reader is referred to the literature.

With negligible exceptions, Table 2A shows that the prevention or inhibition of glycolysis [by omission of glucose, by the addition of 2-deoxy-glucose or by addition of iodo acetic acid (IAA)] failed to affect neuronal retention. This is also true for the inhibition of oxidative phosphorylation [by omission of oxygen or by the addition of dinitrophenol (DNP) or cyanide; Table 2B]. However, a clear impairment of "retention" was observed, whenever both, glycolysis and oxidative phosphorylation, were inhibited (Table 2C). — Any impairment of retention may well involve a decrease of the normal sodium gradient which would a) reduce inward transport and b) increase outward transport.

Wakade and Furchgott (1968) also carried out some experiments without inhibition of MAO. Omission of oxygen increased the retention of

[3]H-noradrenaline, an effect which the authors ascribed to inhibition of MAO.

The energy requirements for the efflux of noradrenaline

Paton (1973) loaded rabbit atria (MAO and COMT inhibited) with ^{3}H-noradrenaline, washed them out for 60 min with amine-free solution and then omitted oxygen and glucose (and added 10 mmol/l Na-azide to the medium). This induced an about five-fold increase in the efflux of tritium (which, under these experimental conditions, represents mainly the ^{3}H-amine). 30 µmol/l cocaine reduced this efflux without abolishing it. Hence, the simultaneous inhibiton of glycolysis and oxidative phosphorylation induced a pronounced outward transport of ^{3}H-noradrenaline.

Schömig et al. (1987) measured the efflux of endogenous noradrenaline and DOPEG (HPLC with electrochemical detection) from rat hearts perfused according to the Langendorff technique. It should be noted that the basal efflux of both compounds was close to the limit of the sensitivity of the method. Hence, decreases or minor increases of the rates of efflux could not be measured with accuracy. Neither hypoxia (i.e., bubbling of the solutions with $N_2 + CO_2$) nor anoxia (i.e., in the additional presence of 0.5 mmol/l dithionite) for 60 min caused any measurable increase in the efflux of noradrenaline, while the efflux of DOPEG became immeasurable. Glucose deprivation (also for 60 min), on the other hand, induced an efflux of noradrenaline. However, hypoxia plus glucose deprivation (and even more so anoxia plus glucose deprivation) elicited a very pronounced efflux of noradrenaline. The combined effect of anoxia plus glucose deprivation was not altered by the absence of calcium from the perfusion medium. Moreover, it was greatly reduced by the presence of 100 nmol/l desipramine. Hence, the combination of hypoxia (or anoxia) with glucose deprivation induces an outward transport of noradrenaline, but anoxia alone is unable to do so. The combination of 1 mmol/l cyanide with glucose deprivation achieved qualitatively similar results (Schömig et al., 1987).

The energy requirements for the efflux of ^{3}H-noradrenaline and its ^{3}H-metabolites from the rat vas deferens (Russ et al., 1991)

The rat vas deferens (COMT inhibited) was selected for this study, since it can be assumed that, because of the very heterogeneous distribution of the ^{3}H-amine, most of the ^{3}H-amine is stored in varicosities that are well supplied with oxygen and glucose.

Ouabain is well known to induce an outward transport of ^{3}H-noradrenaline, since inhibition of the sodium pump increases axoplasmic Na^+ (Paton, 1973; Stute and Trendelenburg, 1984; Sweadner, 1985). It should be noted that ouabain does not inhibit that vesicular ATPase that drives the proton pump responsible for the vesicular uptake of noradrenaline (Apps et al., 1980).

1 mmol/l ouabain induced a desipramine-sensitive efflux of ^{3}H-noradrenaline, i.e., it induced outward transport of the ^{3}H-amine. This was accompanied by a decrease of the efflux of the deaminated ^{3}H-metabolites (^{3}H-DOPEG+^{3}H-DOMA), as predicted by model calculations for compounds that have no other effects within the varicosity (Schömig and Trendelenburg, 1987).

Glucose deprivation likewise induced a desipramine-sensitive outward transport. However, the efflux of deaminated ^{3}H-metabolites increased. This is evidence for an increased net loss of ^{3}H-noradrenaline from the storage vasicles, presumably because of lack of the ATP required by the vesicular ATPase. Surprisingly, this increase of the efflux of deaminated ^{3}H-metabolites was entirely due to an increase in the efflux of ^{3}H-DOMA. It is likely that omission of glucose impairs the formation of NADPH, the cofactor of the aldehyde reductase responsible for the formation of DOPEG.

Hypoxia (induced by bubbling the media with $N_2 + CO_2$) had little effect on the efflux of ^{3}H-noradrenaline, but it quickly reduced the efflux of ^{3}H-DOPEG and ^{3}H-DOMA, by inhibiting the activity of neuronal MAO.

The combination of hypoxia with glucose deprivation induced an efflux of ^{3}H-noradrenaline that was much higher than the sum of efflux increases induced by either procedure alone. Thus, a combination of even a mild degree of hypoxia plus glucose deprivation is able to induce pronounced outward transport of noradrenaline. These results are similar to those shown in Table 2.

Hypoxia in atria may have a further consequence, namely an increased formation of lactate in muscle cells and neurones. Hence, hypoxia (of the vas deferens) was combined with the addition of 20 mmol/l lactic acid (pH of medium reduced from 7.5 to 7.0). While hypoxia alone caused very little desipramine-sensitive outward transport of ^{3}H-noradrenaline (see above), the presence of lactic acid enhanced it.

Lactate formation in vasa deferentia and atria of the rat, before and during hypoxia

Under aerobic conditions lactate formation (measured as efflux of lactate into medium) was higher in atria than in vasa deferentia. In both tissues lactate formation rose steeply on bubbling the medium with $N_2 + CO_2$.

Conclusions

The evidence summarized here supports the view that the high rates of outward transport of ^{3}H-noradrenaline, observed for atria (Schömig et al.,

1989), may well be due to the fact that, as a consequence of the low density of innervation, the exogenous amine is stored in varicosities the metabolic requirements of which are inadequately met. In atria, hypoxia and lactate formation (together with a slight deficiency in the supply of glucose) are able to induce an outward transport of the amine. This in turn means that, in experiments with incubated organs, one can expect not only a heterogeneity for the neuronal distribution of ^{3}H-noradrenaline, one must expect as well a gradient with respect to oxygen (and possibly glucose). These results also indicate that there is no need to claim "organ differences" (vas deferens vs. atria) for adrenergic varicosities; it is likely that the basic properties of adrenergic varicosities are the same in a variety of peripheral organs.

Acknowledgements

The authors are indebted to the Deutsche Forschungsgemeinschaft (SFB 176) and to the Dr. Robert Pfleger-Stiftung for their support of this study. The technical help of M. Fischer and M. Babl is gratefully acknowledged.

References

Apps DK, Pryde JG, Sutton R, Phillips JH (1980) Inhibition of adenosine triphosphatase, 5-hydroxytryptamine transport and proton translocation activities of resealed chromaffin-granule "ghosts". Biochem J 190: 273–282

Azevedo I, Moura D, Trendelenburg U (1990) Autoradiographic study of the rat vas deferens incubated with ^{3}H-noradrenaline. Naunyn-Schmiedebergs Arch Pharmacol 342: 245–248

Hamberger B (1967) Reserpine-resistant uptake of catecholamines in isolated tissues of the rat. Acta Physiol Scand 71[Suppl 295]: 1–56

Hermann W, Graefe K-H (1977) Relationship between the uptake of ^{3}H-metaraminol and the density of adrenergic innervation in isolated rat tissues. Naunyn-Schmiedebergs Arch Pharmacol 296: 99–110

Paton DM (1968) Cation and metabolic requirements for retention of metaraminol by rat uterine horn. Br J Pharmacol 33: 277–286

Paton DM (1972) Metabolic requirements for the uptake of noradrenaline by isolated atria and vas deferens of the rabbit. Pharmacology 7: 78–88

Paton DM (1973) Mechanism of efflux of noradrenaline from adrenergic nerves in rabbit atria. Br J Pharmacol 49: 614–627

Russ H, Schömig E, Trendelenburg U (1991) The energy requirements for the efflux of noradrenaline from adrenergic varicosities. Naunyn-Schmiedebergs Arch Pharmacol 344: 286–296

Schömig E, Trendelenburg U (1987) Simulation of outward transport of neuronal ^{3}H-noradrenaline with the help of a two-compartment model. Naunyn-Schmiedebergs Arch Pharmacol 336: 631–640

Schömig A, Fischer S, Kurz T, Richardt G, Schömig E (1987) Nonexocytotic release of endogenous noradrenaline in the ischemic and anoxic rat heart: mechanism and metabolic requirements. Circ Res 60: 194–205

Schömig E, Fischer P, Schönfeld C-L, Trendelenburg U (1989) The extent of neuronal re-uptake of ^{3}H-noradrenaline in isolated vasa deferentia and atria of the rat. Naunyn-Schmiedebergs Arch Pharmacol 340: 502–508

Schömig E, Schönfeld C-L, Halbrügge T, Graefe K-H, Trendelenburg U (1990) The heterogeneity of the neuronal distribution of exogenous noradrenaline in the rat vas deferens. Naunyn-Schmiedebergs Arch Pharmacol 342: 160–170

Schömig E, Trendelenburg U, Azevedo I, Moura D (1991) The steady-state concentration gradient for ^{3}H-noradrenaline generated by uptake$_1$ in the extracellular space of the rat vas deferens incubated with this amine. Naunyn-Schmiedebergs Arch Pharmacol 344: 41–46

Stute N, Trendelenburg U (1984) The outward transport of axoplasmic noradrenaline induced by a rise of the sodium concentration in the adrenergic nerve endings of the rat vas deferens. Naunyn-Schmiedebergs Arch Pharmacol 327: 124–132

Sweadner KJ (1985) Ouabain-evoked norepinephrine release from intact rat sympathetic neurons: evidence for carrier-mediated release. J Neurosci 5: 2397–2406

Wakade AR, Furchgott RF (1968) Metabolic requirements for the uptake and storage of norepinephrine by the isolated left atrium of the guinea pig. J Pharmacol Exp Ther 163: 123–135

Authors' address: Prof. Dr. U. Trendelenburg, Department of Pharmacology and Toxicology, University of Würzburg, Versbacher Strasse 9, D-W-8700 Würzburg, Federal Republic of Germany

J Neural Transm (1991) [Suppl] 34: 27–35

The effects of type 1 diabetes mellitus and of tobacco smoke on dissipation of catecholamines in pulmonary endothelial cells — a non-neuronal site of uptake$_1$

L. J. Bryan-Lluka and **G. J. McKee**

Department of Physiology and Pharmacology, The University of Queensland,
Queensland, Australia

Summary. The effects of Type 1 diabetes mellitus and of exposure to mainstream cigarette smoke on noradrenaline (NA) uptake and its subsequent metabolism by catechol-O-methyltransferase (COMT) and monoamine oxidase (MAO) in the perfused lungs of rats were examined. In diabetic (streptozotocin-treated) rats, there was an increase in the metabolic clearance of NA in the lungs, and this appeared to be due to the observed increase in NA uptake. During acute exposure of rat lungs to cigarette smoke, there was again an increase in the metabolic clearance of NA, and this was not due to any increase in the activity of either COMT or MAO, but we have not yet investigated whether it is due to an increase in NA uptake. After prolonged exposure of rats to cigarette smoke (daily for 3 months), NA uptake in the lungs was increased, but there was no change in the activity of either COMT or MAO. The results suggest that increased pulmonary clearance may reduce the elevated plasma catecholamine levels that have been described both in patients with Type 1 diabetes mellitus and in cigarette smokers.

Introduction

The lungs are an important site of removal of noradrenaline (NA) from the circulation in various species, including rat (Hughes et al., 1969; Nicholas et al., 1974), rabbit (Gillis and Iwasawa, 1972) and man (Gillis et al., 1972). Uptake of NA occurs into endothelial cells of the capillaries and post-capillary venules and the NA is then metabolized by monoamine oxidase (MAO) and catechol-O-methyltransferase (COMT) (Hughes et al., 1969; Nicholas et al., 1974). The uptake process that transports NA into the endothelial cells has, until recently, been considered to have some of the properties of Uptake$_1$ (the process that transports NA into noradrenergic neurones) and some properties of Uptake$_2$ (or extraneuronal uptake) (Nicholas et al., 1974). However, we have recently shown that uptake of catecholamines into pulmonary endothelial cells, at least in the rat, does, in fact, occur by Uptake$_1$ (Bryan et al., 1988), and hence these cells represent a non-neuronal site of Uptake$_1$.

These studies have now been extended to examine whether the uptake and/or metabolism of NA in pulmonary endothelial cells are altered in rat models of two situations in which increased levels of circulating catecholamines have been reported, *viz.* Type 1 diabetes mellitus (Christensen, 1974) and cigarette smoking (Cryer et al., 1976). Whilst the elevated plasma catecholamine levels appear to be at least partly due to increased adrenal and/or neuronal release in both of these situations, the aim of this study was to determine whether any changes in pulmonary clearance of NA also contribute to or counteract the elevated plasma catecholamine levels.

Materials and methods

Rat model of type 1 diabetes mellitus

Male, specific pathogen free (SPF) Wistar rats, age 4 weeks, were starved overnight, divided into 4 weight-matched groups and treated 29 days prior to sacrifice and removal of the lungs for perfusion (see below). The rats in the diabetic group were given 2 ml/kg normal saline (N.S., 154 mM NaCl) i.p. and 15 min later 75 mg/kg streptozotocin (STZ; in 0.05 M citrate buffer pH 4.5) i.v. to induce damage to the pancreatic β-cells (Rerup, 1970). The rats in the nicotinamide protected group were given 1 g/kg nicotinamide (in N.S.) i.p. to protect the pancreatic β-cells from the lethal actions of STZ (Rerup, 1970) and 15 min later 75 mg/kg STZ i.v., so that any non-specific toxic effects of STZ could be detected. The rats in the nicotinamide control group were included to determine whether nicotinamide itself had any effects and were given 1 g/kg nicotinamide i.p. and 15 min later 1 ml/kg citrate buffer i.v. The rats in the vehicle control group were given 2 ml/kg N.S. i.p. and 15 min later 1 ml/kg citrate buffer i.v. At sacrifice, the body weight gains and the plasma glucose concentrations were not significantly different between the nicotinamide controls (199 ± 8 g, 8.4 ± 0.4 mM, respectively, n = 14) and either the vehicle controls (186 ± 8 g, 8.6 ± 0.4 mM, n = 15) or the nicotinamide protected rats (188 ± 7 g, 8.2 ± 0.2 mM, n = 16). However, the weight gain was significantly less (P < 0.001) and the plasma glucose concentration was significantly greater (P < 0.001) for the diabetic rats (90 ± 11 g, 28.2 ± 1.0 mM, n = 15) than for the nicotinamide protected rats. Rats were considered to be diabetic and included in the study only if the plasma glucose concentration was 20 mM or greater.

Rat model of mainstream cigarette smoke exposure

Mainstream cigarette smoke from standard cigarettes, each containing 1.1 mg nicotine and 13 mg total particulate matter, was generated at the rate of 1 puff of 2 sec duration and 38 ml volume per min, and diluted as described by Griffith and Hancock (1985). In experiments on direct exposure of lungs to cigarette smoke, lungs from male, SPF Wistar rats, age 6–7 weeks, were isolated and perfused (see below) and, from the 5th min of perfusion, were ventilated with diluted cigarette smoke via the tracheal cannula (Griffith and Hancock, 1985). For sham controls, the same system was used except that the ventilation was with air. In experiments on prolonged exposure of rats to cigarette smoke, male SPF Wistar rats, age 5 weeks, were commenced on 10 min exposure to mainstream cigarette smoke (smoke exposure group) or air (sham controls) (Griffith and Standafer, 1985) daily for 3 months prior to sacrifice and perfusion of the lungs.

Lung perfusion experiments

Rats from each of the experimental groups (see above) were sacrificed and the lungs were isolated, perfused and ventilated as described by Bryan-Lluka and O'Donnell (1991).

In experiments to determine the *metabolism* of NA, the lungs were initially perfused via the pulmonary artery at 37°C for 15 min with Krebs solution (containing in mM: NaCl 118, KCl 4.7, $CaCl_2 . 2H_2O$ 2.5, KH_2PO_4 1.2, $MgSO_4 . 7H_2O$ 1.2, $NaHCO_3$ 25.0, glucose 11.7, ascorbic acid 0.57 and Na_2EDTA 0.040) and 5% bovine serum albumin at 10 ml/min. The lungs were then perfused for 20 min with aerated (95% O_2 and 5% CO_2) Krebs solution containing 1 nM ^3H-(-)-NA. Samples of venous effluent and lung homogenate were obtained and analyzed as described by Bryan-Lluka and O'Donnell (1991) to separate NA and its metabolites, dihydoxyphenylethyleneglycol (DOPEG), dihydroxymandelic acid (DOMA), normetanephrine (NMN) and the two O-methylated deaminated metabolites (OMDA) in one further fraction. The tissue to medium ratio of NA (T/M_{NA}) in the lungs was calculated from the NA content of the lungs at the end of the experiment, the NA concentration (1 nM) and the extracellular space (ECS) of the lungs measured in separate experiments (0.461 ml/g).

Values for the *activity of COMT* (k_{COMT}) were determined in experiments carried out as described above, but with MAO inhibited by pretreatment of the rats with 75 mg/kg pargyline i.p. 18 h and 2 h prior to sacrifice. Values of k_{COMT} were calculated as the steady-state rate of NMN formation divided by the product of T/M_{NA} and the NA concentration (1 nM).

Values for the *activity of MAO* (k_{MAO}) were determined in experiments carried out as described above, but with COMT inhibited by inclusion of 10 μM U-0521 (3', 4'-dihydroxy-2-methylpropiophenone) in the perfusion solution throughout the experiments. Values of k_{MAO} were calculated as for k_{COMT}, but using steady-state rates of formation of DOPEG + DOMA.

In experiments to determine the *uptake* of NA in the lungs, MAO and COMT were inhibited and the lungs were perfused as described above, except that 2 nM ^3H-NA was perfused for only 2 min to measure initial rates of uptake. ^{14}C-Sorbitol (100 μM) was also included to determine the ECS of the lungs and hence correct rates of NA uptake for distribution in the ECS.

Calculation of results

Results are expressed as arithmetic means ± s.e., except k_{COMT} and k_{MAO} values which are expressed as geometric means with 95% confidence limits (C.L.). The significance of differences between mean values was assessed by Student's *t*-test.

Results

Effects of type 1 diabetes mellitus on uptake and metabolism of noradrenaline in rat lungs

Lungs from vehicle control, nicotinamide control, nicotinamide protected and diabetic groups of rats were perfused with NA (see Materials and methods) and the metabolites in the venous effluent and lung homogenates were measured. There were no significant differences in the formation of any of the metabolites between the nicotinamide control group (Fig. 1B)

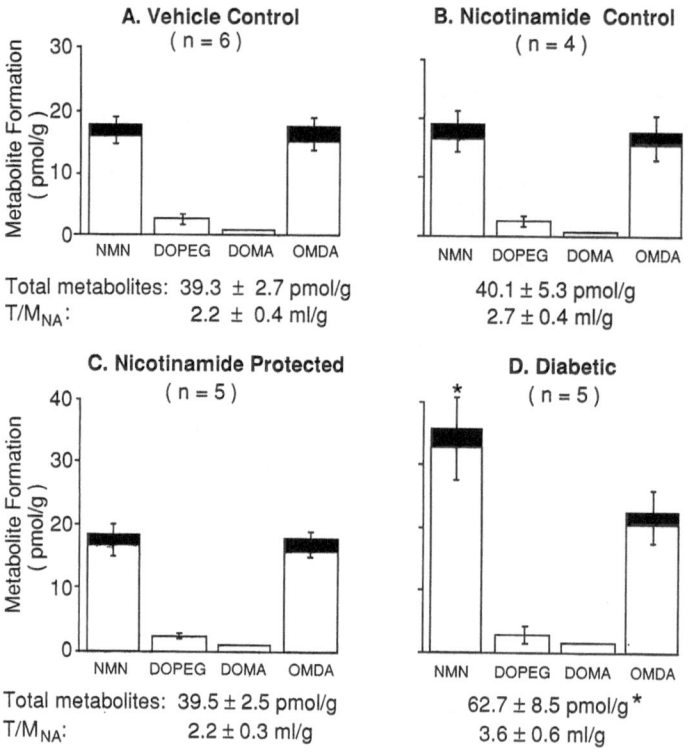

Fig. 1. Comparison of metabolite formation in lungs perfused with 1 nM ^{3}H-NA for 20 min from **A** vehicle control, **B** nicotinamide control, **C** nicotinamide protected and **D** diabetic groups of rats. Details of treatments and abbreviations of metabolites are in Materials and methods. Neither COMT nor MAO was inhibited. The metabolites measured in the venous effluent (□) and in the lung homogenate at the end of the experiment (■), total metabolite formation and T/M$_{NA}$ are shown as means ± s.e. for n rats in each group. *Significant difference between corresponding values for diabetic and nicotinamide protected groups, P < 0.05

and either the vehicle control (Fig. 1A) or the nicotinamide protected (Fig. 1C) groups. In the diabetic group (Fig. 1D), there was a significant increase in the total formation of metabolites by 59%, due to a significant increase in NMN formation by 96%, compared with the nicotinamide protected group (Fig. 1C). In a further series of experiments, the rate of uptake of NA was measured in lungs from each of the 4 groups of rats (see Materials and methods). The rate of NA uptake for the nicotinamide control group (4.01 ± 0.50 pmol/g/min, n = 5) was not significantly different from that for either the vehicle control (4.50 ± 0.35 pmol/g/min, n = 5) or the nicotinamide protected (4.13 ± 0.36 pmol/g/min, n = 6) groups. However, the rate of NA uptake for the diabetic group (6.50 ± 0.94 pmol/g/min, n = 6) was significantly increased (P < 0.05) by 57%, compared with the nicotinamide protected group.

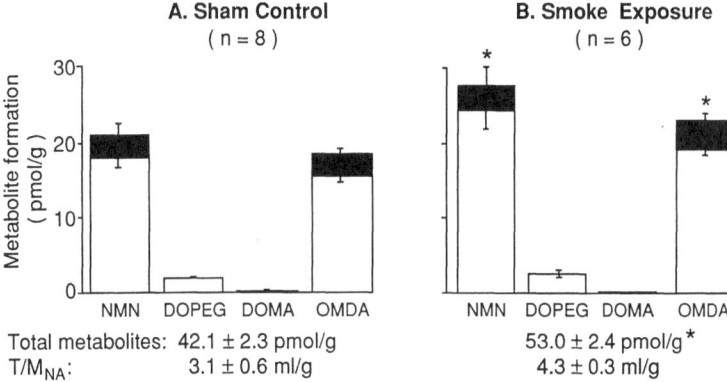

Fig. 2. Comparison of metabolite formation in lungs perfused with 1 nM ^{3}H-NA for 20 min for **A** sham controls and **B** smoke exposure (lungs exposed to cigarette smoke). Details and abbreviations of metabolites are in Materials and methods. Neither COMT nor MAO was inhibited. The metabolites measured in the venous effluent (□) and in the lung homogenate at the end of the experiment (■), total metabolite formation and T/M_{NA} are shown as means ± s.e. for n rats in each group. *Significant difference between corresponding values for smoke exposure and sham control groups, $P < 0.05$

Effects of direct exposure of rat lungs to mainstream cigarette smoke on uptake and metabolism of noradrenaline

Rat lungs were perfused with NA and exposed to either air (sham controls) or cigarette smoke (smoke exposure group) (see Materials and methods). The NA and metabolites in the venous effluent and lung homogenate were measured. In the lungs with smoke exposure (Fig. 2B), there was a significant increase in the total formation of metabolites by 26%, due to significant increases in the formation of NMN by 31% and OMDA by 22%, compared with the sham controls (Fig. 2A). In further experiments, the activity of COMT and of MAO were determined (see Materials and methods) in sham control and smoke exposed lungs. Values of k_{COMT} and k_{MAO} for the smoke exposure group (0.396 min^{-1}, 95% C.L. 0.285, 0.551 min^{-1}, n = 4 and 0.193 min^{-1}, 95% C.L. 0.119, 0.311 min^{-1}, n = 4, respectively) were not significantly different from those for the sham controls (0.406 min^{-1}, 95% C.L. 0.229, 0.719 min^{-1}, n = 4 and 0.198 min^{-1}, 95% C.L. 0.129, 0.304 min^{-1}, n = 4, respectively).

Effects of prolonged exposure of rats to mainstream cigarette smoke on uptake and metabolism of noradrenaline in the lungs

After daily exposure of rats to mainstream cigarette smoke (smoke exposure group) or air (sham controls) for 3 months, the lungs were perfused with NA under conditions to determine the rate of NA uptake, the activity of COMT or the activity of MAO (see Materials and methods). The results (Table 1) show that, in the smoke exposure group, there was a

Table 1. Effects of exposure of rats to mainstream cigarette smoke daily for 3 months on uptake and metabolism of noradrenaline (NA) in perfused lungs

Treatment group[a]	Uptake[b]	Metabolism[c]	
	Rate of uptake of NA (pmol/g/min)	k_{COMT} (min^{-1})	k_{MAO} (min^{-1})
Sham control	4.63 ± 0.17 (n = 5)	0.488 (0.381; 0.625) (n = 4)	0.292 (0.200; 0.427) (n = 5)
Smoke exposure	5.90 ± 0.48* (n = 6)	0.510 (0.285; 0.914) (n = 4)	0.291 (0.162; 0.525) (n = 5)

[a] The smoke exposure groups of rats were exposed to mainstream cigarette smoke for 10 min daily for 3 months. The sham control groups were exposed to the same treatment, but without the cigarette smoke.
[b] The lungs of rats were perfused with 2 nM ^{3}H-NA for 2 min. COMT and MAO were inhibited. (Details in Materials and methods.) Data are means ±s.e. for n rats.
[c] The lungs of rats were perfused with 1 nM ^{3}H-NA for 20 min. MAO was inibited in experiments to determine k_{COMT}; COMT was inhibited in experiments to determine k_{MAO}. (Details in Materials and methods.) Data are geometric means with 95% confidence limits for n rats.
* Significant difference between corresponding values for smoke exposure and sham control groups, $P < 0.05$

significant increase in the rate of NA uptake in the lungs by 27%, compared with the sham controls, but there was no change in the activity of either COMT or MAO.

Discussion

Type 1 diabetes mellitus is associated with structural changes in vascular endothelial cells in human patients (Stary, 1966) and in STZ-treated diabetic rats (Moore et al., 1985). Changes have also been described in the metabolic functions of endothelial cells in diabetes mellitus, such as decreased prostacyclin production in systemic vascular endothelial cells in human patients (Johnson et al., 1979). Reports from studies on lungs from STZ-treated diabetic rats have also shown that experimental diabetes mellitus causes a decrease in prostacyclin production (Watts et al., 1982) and in removal of ADP and AMP (Bakhle and Chelliah, 1983), but no change in removal of 5-hydroxytryptamine (5-HT) (Watkins and Rannels, 1982) in the pulmonary circulation. The present study has shown that the metabolic clearance of NA in the lungs of STZ-treated diabetic rats is increased, compared with that in non-diabetic controls, with a marked increase particularly in the formation of the O-methylated metabolite, NMN. The results also show that the increase in NA clearance in the lungs of the diabetic rats appears to be due to an increase in the uptake of NA into the pulmonary endothelial cells. However, the mechanism for the apparent stimulation of Uptake$_1$ of NA into the pulmonary endothelial cells of the diabetic rats is not yet clear, and also experiments have not yet been carried out to determine whether Uptake$_1$ into noradrenergic neurones is also stimulated in diabetes mellitus. The selective increase in metabolism of NA by COMT in the diabetic rats could be due to the combined effects of increased transport of NA into the cells and a decrease in the activity of MAO, which has been reported to occur in human platelets (Mosnaim et al., 1979), but has not yet been investigated in rat lungs.

Cigarette smoking is another condition in which structural changes have been reported to occur in vascular endothelial cells in humans (Asmussen and Kjeldsen, 1975) and in rats (Pittilo et al., 1982). Metabolic changes have been reported in pulmonary endothelial cells during direct ventilation of isolated lungs of rats or hamsters with mainstream cigarette smoke (model of acute effects during active smoking), including reduced inactivation of 5-HT (Karhi et al., 1982) and arachidonic acid (Matintalo and Uotila, 1983). However, exposure of rats or hamsters to mainstream cigarette smoke for 1 or 10 days (model of short-term active smoking) did not affect the pulmonary inactivation of 5-HT (Bakhle et al., 1979) or arachidonic acid (Männisto et al., 1981). In the present study, we investigated whether the dissipation of NA in the pulmonary circulation of rats is affected either by direct exposure of the isolated lungs to mainstream cigarette smoke or by daily exposure of rats to mainstream cigarette smoke for a prolonged period (3 months). The results showed that direct smoke exposure of the lungs resulted in an increase in the metabolic clearance of NA in the lungs, and this was not attributable to any change in the activity of either COMT or MAO, but we have not yet investigated whether it is due to stimulation of uptake of NA into the pulmonary endothelial cells. Exposure of rats to cigarette smoke daily for 3 months resulted in an increase in uptake of NA into the pulmonary endothelial cells, but no change in the activity of either COMT or MAO. We have not yet investigated the mechanism of this apparent stimulation of Uptake$_1$ into the pulmonary endothelial cells after prolonged exposure to cigarette smoke nor the possibility that stimulation of Uptake$_1$ into noradrenergic neurones may also occur.

In conclusion, the study has shown that, in rat models both of Type 1 diabetes mellitus and of acute and prolonged active smoking, there appears to be an increase in the clearance of NA in the pulmonary circulation. If these effects also occur in man, then enhanced pulmonary clearance may reduce the elevated plasma catecholamine levels that have been described both in patients with diabetes mellitus and in cigarette smokers.

Acknowledgements

We would like to thank H. Vuocolo and J. Brown for excellent technical assistance in some of the experiments and Upjohn Pty Ltd. for a gift of U-0521. The support of the research on the effects of tobacco smoke and the supply of the standard cigarettes by the Australian Tobacco Research Foundation are also gratefully acknowledged.

References

Asmussen I, Kjeldsen K (1975) Intimal ultrastructure of human umbilical arteries. Circ Res 36: 579–589

Bakhle YS, Chelliah R (1983) Effect of streptozotocin-induced diabetes on the metabolism of ADP, AMP and adenosine in the pulmonary circulation of rat isolated lung. Diabetologia 24: 455–459

Bakhle YS, Hartiala J, Toivonen H, Uotila P (1979) Effects of cigarette smoke on the metabolism of vasoactive hormones in rat isolated lungs. Br J Pharmacol 65: 495–499

Bryan-Lluka LJ, O'Donnell SR (1991) Isolated perfused lungs of guinea-pig, in contrast with rat, lack an uptake process for noradrenaline. Pulm Pharmacol 4: 146–150

Bryan LJ, O'Donnell SR, Westwood NN (1988) The uptake process for catecholamines in endothelial cells in rat perfused lungs is the same as Uptake$_1$ in noradrenergic neurones. Br J Pharmacol 95: 539P

Christensen NJ (1974) Plasma norepinephrine and epinephrine in untreated diabetics, during fasting and after insulin administration. Diabetes 23: 1–8

Cryer, PE, Haymond MW, Santiago JV, Shah SD (1976) Norepinephrine and epinephrine release and adrenergic mediation of smoking-associated hemodynamic and metabolic events. N Engl J Med 295: 573–577

Gillis CN, Iwasawa Y (1972) Technique for measurement of norepinephrine and 5-hydroxytryptamine uptake by rabbit lung. J Appl Physiol 33: 404–408

Gillis CN, Greene NM, Cronau LH, Hammond GL (1972) Pulmonary extraction of 5-hydroxytryptamine and norepinephrine before and after cardiopulmonary bypass in man. Circ Res 30: 666–674

Griffith RB, Hancock R (1985) Simultaneous mainstream-sidestream smoke exposure systems I. Equipment and procedures. Toxicology 34: 123–138

Griffith RB, Standafer S (1985) Simultaneous mainstream-sidestream smoke exposure systems II. The rat exposure system. Toxicology 35: 13–24

Hughes J, Gillis CN, Bloom FE (1969) The uptake and disposition of *dl*-norepinephrine in perfused rat lung. J Pharmacol Exp Ther 169: 237–248

Johnson M, Harrison HE, Raftery AT, Elder JB (1979) Vascular prostacyclin may be reduced in diabetes in man. Lancet i: 325–326

Karhi T, Rantala A, Toivonen H (1982) Pulmonary inactivation of 5-hydroxytryptamine is decreased during cigarette smoke ventilation of rat isolated lungs. Br J Pharmacol 77: 245–248

Männisto J, Toivonen H, Hartiala J, Bakhle YS, Uotila P (1981) The effect of cigarette smoke on the metabolism of arachidonic acid in isolated hamster lungs. Prostaglandins 22: 195–204

Matintalo M, Uotila P (1983) The effect of low-, medium-, and high-tar cigarette smoke on the fate of arachidonic acid in isolated hamster lungs. Acta Pharmacol Toxicol 53: 280–287

Moore SA, Bohlen HG, Miller BG, Evan AP (1985) Cellular and vessel wall morphology of cerebral cortical arterioles after short-term diabetes in adult rats. Blood Vessels 22: 265–277

Mosnaim AD, Wolf ME, Huprikar S, Singh SP, Zeller EA (1979) Reduced monoamine oxidase activity in blood platelets from insulin-dependent diabetic subjects. Diabetes 28: 455–456

Nicholas TE, Strum JM, Angelo LS, Junod AF (1974) Site and mechanism of uptake of ^{3}H-l-norepinephrine by isolated perfused rat lungs. Circ Res 35: 670–680

Pittilo RM, Mackie IJ, Rowles PM, Machin SJ, Woolf N (1982) Effects of cigarette smoking on the ultrastructure of rat thoracic aorta and its ability to produce prostacyclin. Thromb Haemostas 48: 173–176

Rerup CC (1970) Drugs producing diabetes through damage of the insulin secreting cells. Pharmacol Rev 22: 485–518

Stary HC (1966) Disease of small blood vessels in diabetes mellitus. Am J Med Sci 252: 357–374

Watkins CA, Rannels DE (1982) Effect of diabetes on metabolism of 5-hydroxy-tryptamine by rat lungs perfused *in situ*. Am Rev Resp Dis 126: 175–177

Watts IS, Zakrzewski JT, Bakhle YS (1982) Altered prostaglandin synthesis in isolated lungs of rats with streptozotocin-induced diabetes. Thromb Res 28: 333–342

Authors' address: Dr. L. J. Bryan-Lluka, Department of Physiology and Pharmacology, The University of Queensland, Queensland 4072, Australia

J Neural Transm (1991) [Suppl] 34: 37–42

Distribution of extraneuronal uptake$_1$ in reproductive tissues: studies on cells in culture

I. S. de la Lande[1], V. Marino[2], T. Lavranos[3], J. A. Kennedy, D. A. S. Parker[2], and **R. F. Seamark[3]**

Departments of [1]Clinical and Experimental Pharmacology, [2]Dentistry, and [3]Obstetrics and Gynaecology, University of Adelaide, Adelaide, Australia

Summary. Cultures of stromal cells from pregnant mouse uterus, and an FL cell line derived from human amnion, displayed significant capacities to O-methylate noradrenaline. O-methylation was inhibited in the stromal cells by uptake$_1$-inhibitors, and in the FL cell line by uptake$_2$ inhibitors. These findings are discussed in terms of the distribution and possible functional importance of catecholamine metabolising systems in the female reproductive system.

Introduction

Earlier studies on noradrenaline (NA) metabolism in rabbit uterine endometrium, a tissue devoid of sympathetic innervation, and on the sympathetically-denervated dental pulp, revealed the presence of an extraneuronal uptake system which, like uptake$_2$, was linked with O-methylation, yet had the characteristics of uptake$_1$. The characteristics included (a) sensitivity to inhibitors of uptake$_1$ but not uptake$_2$, (b) high affinity for unlabelled, NA, as indicated by IC-50s for inhibition of ^{3}HNA uptake in the 1–2 µM range, and (c) Na$^+$ dependency (Kennedy and de la Lande, 1986, 1987; Parker et al., 1987; de la Lande et al., 1989, 1990). An initial difficulty facing identity of the uptake system in the endometrium with uptake$_1$ was a high apparent Km for NA (76 µM; Kennedy and de la Lande, 1987). The difficulty was resolved with the finding that a second uptake process, characterised by sensitivity to hydrocortisone, became important in the upper range of concentrations of ^{3}HNA used to estimate its Km (de la Lande et al., 1990). When the second uptake was eliminated and the affinity measured in terms of the IC50, the value of the Km obtained (1.6 µM) was compatible with uptake$_1$.

In addition to the endometrium and dental pulp, the presence of extraneuronal uptake$_1$ in rabbit gingiva and nasal mucosa, and in pregnant rat endometrium and placenta, is suggested by the sensitivity of NA O-methylation to cocaine (Parker et al., 1987; de la Lande et al., 1987; Kennedy and de la Lande, 1988). In a current study, we have extended an investigation of the distribution of extraneuronal uptake$_1$ in tissues and cells to include cells grown in culture. Epithelial cells and fibroblasts were selected since glandular epithelium has been identified as the site of cocaine-sensitive uptake in the rabbit endometrium (Kennedy and de la Lande, 1986), and fibroblast-like cells as the sites in dental pulp (Parker et al., 1987). An added factor was the availability of a number of cell cultures from the Department of Obstetrics and Gynecology, University of Adelaide. These were being used to study the in vitro development of the mouse blastocyst. In view of the endometrial location of extraneuronal uptake$_1$, it was of interest to ascertain whether this uptake was present in cells which supported blastocyst development.

Methods

Primary cell cultures

Stromal and epithelial cells from pregnant mouse endometrium were prepared by the method of Robertson and Seamark (1989) and cultured under the conditions described by Lavranos and Seamark (1989).

Established cell lines

These comprised L-cells, derived from mouse fibroblasts (NCTC clone; ATCC catalogue 1988) and FL cells, derived from human amnion epithelial cells (ATCC catalogue, 1988).

Culture medium

Cells were cultured in Minimum Essential Medium, supplemented as described by Lavranos and Seamark (1989). At all stages, including subsequent incubation with ^{3}HNA, the cell cultures were in an atmosphere of air-5% CO_2 at 37°C.

Further treatment

The cells were seeded into 12-well multi-dishes in a density of approximately 500,000 cells per well. Usually the cells confluenced into a monolayer within 24 h. After the medium had been replaced twice at 15 min intervals with fresh medium (0.5 ml), ^{3}HNA (1.0 µM) was added and the incubation continued for 30 min. Drugs, including the COMT inhibitor, U-0521, were added 15 min before the ^{3}HNA. At the end of incubation, the cells were rinsed briefly (5 s), after which the contents of ^{3}HNA and its metabolites were assayed by the column chromatographic method of Graefe et al. (1973). The methods of extracting the cells and preparation of media and extracts for assay were identical to those used for tissue slices (see Kennedy and de la Lande, 1986). It should be noted that, in the assay, crossover of NA into the OMDA and DOMA fractions is considerably higher than into the NMN and DOPEG fractions. High crossover precluded reliable estimates of DOMA in most experiments.

Drugs and chemicals

2, 5, 6-^{3}H($-$) noradrenaline, specific activity 44 Ci/mmol (New England Nuclear); ($-$) noradrenaline bitartrate, desipramine hydrochloride (Sigma); cocaine hydrochloride (Macfarlane Smith); hydrocortisone hemisuccinate and 3, 4-dihydroxy-2-methylpropriophenone (U-0521) (Upjohn); 3 O-methylisoprenaline (Boehringer).

Abbreviations

a) NA metabolites: *OMDA* o-methylated deaminated fraction, *NMN* normetanephrine, *DOMA* dihydroxy-mandelic acid, *DOPEG* dihydroxyphenylglycol
b) *COMT* catechol-O-methyltransferase

Results

Uterine cells

In each of 8 primary cultures, mouse stromal cells metabolised ^{3}HNA (1.0 μM) to O-methylated and deaminated products; metabolism in the uterine epithelial cells was poor by comparison (Table 1). There was little evidence of accumulation of ^{3}HNA by either cell-type. In the stromal cells, NMN was a major metabolite. NMN formation was virtually abolished by cocaine (30 μM) and desipramine (3 μM), but was relatively unaffected by hydrocortisone (100 μM) and 3-O-methylisoprenaline (20 μM) (Table 2). Effects of cocaine and desipramine on DOPEG formation resembled those on NMN formation, but were less pronounced (data not shown). The quantities of NMN formed by the epithelial cells were too small to justify analysis of drug effects.

Table 1. Metabolism of ^{3}HNA in cell cultures[a]

| Origin | Cell type | ^{3}H Metabolite formation | | | |
		NMN	OMDA	DOPEG	^{3}HNA cell content
Mouse endometrium	Stromal (fibroblast) n = 8	2.70 ± 0.90	3.60 ± 0.90	1.00 ± 0.30	0.29 ± 0.09
	epithelial n = 5	0.28 ± 0.10	0.73 ± 0.50	0.18 ± 0.13	0.24 ± 0.09
FL cells	ex amnion n = 8	1.40 ± 0.36	0.57 ± 0.27	0.03 ± 0.02	0.22 ± 0.10

[a] 3HNA (1.0 μM) for 30 min; data in pmol/well

Table 2. Effects of uptake inhibitors on formation[a] of NMN

Drug	Stromal cells	FL cells
Control	3.30 ± 1.40	1.3 ± 0.3
Cocaine (30 µM)	0.09 ± 0.05	1.3 ± 0.3
Control	1.40 ± 0.60	2.1 ± 0.6
Desipramine (3 µM)	0.08 ± 0.06	1.5 ± 0.5
Control	1.00 ± 0.30	1.4 ± 0.6
Hydrocortisone (100 µM)	1.00 ± 0.20	0.4 ± 0.2
Control	1.50 ± 0.40	2.1 ± 0.6
Meoiso (20 µM)	1.00 ± 0.30	0.5 ± 0.1

[a] 3 HNA (1.0µM) for 30 min, data in pmol/well (n = 4); *Meoiso* O-methylisoprenaline

FL cells

^{3}HNMN was the principal metabolite, ^{3}HDOPEG being almost undetectable (Table 1). The cellular content of ^{3}HNA was small compared with metabolite formation but increased dramatically (from 0.22 ± 1.10 to 1.6 ± 0.2 pmol/well, n = 8), following inhibition of ^{3}HNMN formation by the COMT inhibitor, U-0521 (30 µM). ^{3}HNMN formation was unaffected by cocaine (30 µM) and desipramine (3 µM), but was inhibited quite strongly by hydrocortisone (100 µM) and 3-O-methylisoprenaline (20 µM) (Table 2).

L-cells

There was no evidence of metabolism or uptake of ^{3}HNA by these cells.

Discussion

Although only two of the four cell types metabolised NA to an extent which justified their further study, in both metabolism appeared to be linked to an uptake process. In one (mouse stromal cells), sensitivity of O-methylation to cocaine and desipramine implies the presence of uptake$_1$. In the other (FL cells), inhibition of O-methylation by hydrocortisone and 3-O-methylisoprenaline implies the presence of uptake$_2$. The ease of culture of the FL cells has enabled a detailed analysis of its uptake system, the results of which (to be reported) indicate that it is very simlar to uptake in the Caki renal tubular cell line whose identity with uptake$_2$ has been established by Schömig and Schönfeld (1990) (Marino and de la Lande, unpublished observations). Unfortunately, limited availability has precluded further analysis of the uptake system in the stromal cells.

The high O-methylating activity of the mouse stromal cells relative to the epithelial cells was unexpected, since in the rabbit endometrium, glandular epithelial cells and not the stromal cells are the site of cocaine-sensitive uptake (Kennedy and de la Lande, 1986). The difference further emphasizes the diversity of cellular sites of uptake$_1$, particularly when pulmonary endothelium (Nicholas et al., 1974) and the sympathetic neurone are added to the list, and at the same time implies that the cellular site in the same organ may differ in different species.

The present results, by extending the probable distribution of extra-neuronal uptake$_1$ to include uterine endometria of three species (mouse, rabbit and rat) draw attention to the possible role of this uptake in reproductive function. In the rat endometrium, cocaine-sensitive O-methylation increases rapidly to day 5 of gestation, after which it progressively declines (Kennedy and de la Lande, 1988). This time course suggests that the role of the uptake$_1$ system may be more important in uterine function during early rather than late pregnancy and prompted us to consider the possibility, mentioned in the Introduction, that cells which maintain the blastocyst in culture may also possess this uptake system. However the possibility can be discounted, since the cells which had this ability were the mouse stromal and epithelial cells and the L-cell line, yet only the stromal cells diplayed significant O-methylating activity. It seems that the presence of extra-neuronal uptake$_1$ is unimportant to early embryo development in vitro. Of course, this may not be the case when noradrenaline is present in the culture medium, having in mind that the endometrial NA metabolizing system is strategically located to minimize exposure of the developing blastocyst to maternal catecholamines.

Since endometria of rabbit and rat do not possess a sympathetic innervation (Kennedy and de la Lande, 1986; Garfield, 1986) it is probable that the endogenous catecholamines derived from the maternal circulation, and possibly also from the foetus, are the endogenous substrates for extra-neuronal uptake$_1$. However catechol oestrogens as well as catecholamines are endogenous substrates for COMT. Since there is evidence that, in the pig, the blastocyst is a source of catechol oestrogens (Mondschein et al., 1985), it is possible that the two catechols compete for the COMT which is linked with uptake$_1$. Kennedy explored this possibility by examining the effects of the catechol oestrogens on O-methylation of NA in the rabbit endometrium. She included rabbit myometrium in the study since in this tissue metabolism of NA in low concentrations appears to be mediated via uptake$_2$ (Kennedy and de la Lande, 1986). The steroids proved to be potent inhibitors in the myometrium, but were far less potent, by almost two orders of magnitude, in the endometrium (Kennedy, 1991). Although these results point to uptake$_2$ rather than uptake$_1$ as the site of interaction between catecholamine and oestrogen metabolism, there is the intriguing implication that, of the NA metabolising systems which are linked with uptake$_1$ and uptake$_2$, the former is less likely to be perturbed by the high levels of oestrogen occurring in pregnancy. Perhaps this factor accounts for the occurrence of extraneuronal-uptake$_1$ metabolizing systems in organs of female reproduction.

Acknowledgement

It is a pleasure to acknowledge the assistance and inspiration provided over many years by U. Trendelenburg in the studies on catecholamine metabolism carried out by the senior investigator (ISD) at the University of Adelaide.

References

de la Lande IS, Parker DAS, Procter C, Marino V (1987) Cocaine inhibits extraneuronal O-methylation of exogenous noradrenaline in nasal and oral tissues of the rabbit. Life Sci 41: 2463–2468

de la Lande Is, Parker DAS, Marino V (1989) O-methylation of noradrenaline is linked with an uptake process which resembles uptake$_1$ in dental pulp. J Neurochem 52 [Suppl]: 16

de la Lande IS, Marino V, Kennedy JA, Parker DAS (1990) Further evidence that the extraneuronal uptake of noradrenaline in rabbit dental pulp is similar to uptake$_1$. Eur J Pharmacol 183: 1147–1148

Garfield RE (1986) Structural studies on innervation of non pregnant rat uterus. Am J Physiol 251: C41–54

Graefe K-H, Stefano FJE, Langer SZ (1973) Prefential metabolism of $(-)$-^{3}H-noradrenaline through the deaminated glycol in the rat vas deferens. Biochem Pharmacol 22: 1147–1160

Kennedy JA (1991) Effect of catechol oestrogens on extraneuronal metabolism of noradrenaline by rabbit uterine endometrium and myometrium. Naunyn-Schmiedebergs Arch Pharmacol 343: 266–270

Kennedy JA, de la Lande IS (1986) Effect of progesterone on the metabolism of noradrenaline in the rabbit uterine endometrium and myometrium. Naunyn-Schmiedebergs Arch Pharmacol 333: 368–376

Kennedy JA, de la Lande (1987) Characteristics of cocaine-sensitive accumulation and O-methylation of noradrenaline by rabbit endometrium. Naunyn-Schmiedebergs Arch Pharmacol 336: 148–154

Kennedy JA, de la Lande IS (1988) Catecholamine inactivation in uterine tissues; changes in pregnancy in the rat. Clin Exp Physiol Pharmacol 15: 675–680

Lavranos TC, Seamark RF (1989) Addition of steroids to embryo-uterine monolayer co-culture enhances embryo survival and implantation in vitro. Reprod Fertil Dev 1: 41–46

Mondschein JS, Hersey RM, Dey SK, Davis D, Weisz J (1985) Catechol estrogen formation by pig blastocysts: biochemical characterization of estrogen-2/4-hydroxylase and correlation with aromatase activity during the preimplantation period. Endocrinology 117: 2339–2346

Nicholas TE, Strum JM, Angelo IS, Junod AF (1974) Site and mechanism of uptake of ^{3}H-1-noradrenaline by isolated perfused rat lungs. Circ Res 35: 670–680

Parker DAS, de la Lande IS, Proctor C, Marino V, Lam NX, Parker I (1987) Cocaine-sensitive O-methylation of noradrenaline in dental pulp; comparison with the rabbit ear artery. Naunyn-Schmiedebergs Arch Pharmacol 336: 32–39

Robertson SA, Searmark RF (1990) Granulocyte macrophage colony stimulating factor (GM-CSF) in the murine reproductive tract; stimulation by seminal factors. Reprod Fertil Dev 2: 359–368

Schömig E, Schönfeld C-L (1990) Extraneuronal noradrenaline transport (uptake$_2$) in a human cell line. Naunyn-Schmiedebergs Arch Pharmacol 341: 404–410

Authors' address: Prof. I. S. de la Lande, Department of Clinical and Experimental Pharmacology, University of Adelaide, Box 498, G. P. O., Adelaide, South Australia 5001

J Neural Transm (1991) [Suppl] 34: 43–49

Extraneuronal inactivation of noradrenaline in tissue culture

E. Schömig, J. Babin-Ebell, M. Gliese, and **H. Russ**

Department of Pharmacology, University of Würzburg, Federal Republic of Germany

Summary. Corticosterone-sensitive extraneuronal transport (uptake$_2$) and metabolism of noradrenaline was investigated in the clonal human Caki-1 cell line. Caki-1 cells are the first experimental system for uptake$_2$ which is based on a tissue culture technique. Previous experiments with Caki-1 cells opened the possibility of a close relationship between uptake$_2$ and the renal transport system for organic cations (RTOC). The finding that steroids potently inhibit both uptake$_2$ and RTOC further supports this hypothesis. Noradrenaline which has been taken up into Caki-1 cells by uptake$_2$ is metabolized by the intracellular enzymes catechol-O-methyltransferase (COMT) and to a lesser extend by monoamineoxidase (MAO).

Introduction

The neurotransmitter noradrenaline is removed from the extracellular space by active transport processes. Two clearly distinct transport mechanisms have been identified. On the one hand, the desipramine-sensitive transport system (uptake$_1$) has been demonstrated in noradrenergic neurones (for review see Graefe and Bönisch, 1988) and in adrenal medullary cells (Coppeneur et al., 1991). Recent findings indicate that uptake$_1$ exists also in several non-neuronal tissues such as dental pulp (Kennedy and de la Lande, 1987), endometrium (Parker et al., 1987) and lung endothelium (Bryan, 1990). On the other hand, the corticosterone-sensitive transport mechanism (uptake$_2$) has been demonstrated in various extraneuronal tissues such as myocardium and vascular smooth muscle (for review see Trendelenburg, 1988).

The majority of investigations on noradrenaline transport were done on isolated incubated or perfused sympathetically innervated organs. However, the use of clonal cell lines facilitates transmembrane transport studies significantly. Transport experiments with clonal cells are not complicated by the distribution of the substate in the extracellular space, by different types of cells, and by diffusion barriers. Moreover, a clonal cell line may serve as primary material for the molecular characterization of carrier proteins. Clonal PC12 rat phaeochromocytoma cells have been used and studied extensively with respect to functional aspects of uptake$_1$. Most recently, the complementary DNA encoding the human uptake$_1$ carrier has been isolated by an expression cloning strategy on the basis of the SK-N-SH cell line (Pacholczyk et al., 1991). Several attempts were made in order to identify a clonal cell line with uptake$_2$. In 1990 Schömig and Schönfeld reported uptake$_2$ to exist in the human clonal Caki-1 cell line. This paper briefly reviews recent work from our laboratory in this field.

Material and methods

Cell culture

Caki-1 cells (American Type Culture Collection, 1988, ATCC HTB 46) and LLC-PK$_1$ cells (American Type Culture Collection, 1988, ATCC CL 101) were grown at 37°C in a humified atmosphere (5% CO_2) on plastic culture flasks (Falcon 175 cm^2, Becton Dickinson, Heidelberg, FRG). The culture medium was composed of Dulbecco's Modified Eagle Medium (5.6 mmol/l D(+)glucose), 10% fetal calf serum, and 4 mmol/l L-glutamine (all from Gibco, Eggenstein, FRG). The cultures were splitted every 4 days (0.05% trypsin, 5 min, 37°C). For the experiments, the cells were seeded on plastic culture dishes. After 4 days, the cells formed confluent monolayers. Intracellular volume was determined as tritium water space.

Uptake and metabolism experiments

Uptake experiments were performed essentially as described by Schömig and Schönfeld (1990). The cells were preincubated with buffer A (125 mmol/l NaCl, 4.8 mmol/l KCl, 1.2 mmol/l KH$_2$PO$_4$, 1.2 mmol/l Mg$_2$SO$_4$, 1.2 mmol/l CaCl$_2$, 25 mmol/l HEPES. NaOH pH 7.4, 5.6 mmol/l D(+)glucose, 1 mmol/l L(+)ascorbic acid) at 37°C for 20 min. Subsequently, the cells were incubated with ^{3}H-noradrenaline and ^{14}C-tetraethylammonium, respectively. Inhibitors of transport or of metabolizing enzymes were present during the preincubation and during the incubation period. Incubation was stopped by rinsing the cells four times with 3.5 ml ice-cold buffer A. Subsequently the cells were solubilized by 0.4 mol/l perchloric acid. Noradrenaline and its metabolites were separated by column chromatography (Graefe et al., 1973).

Results and discussion

Uptake$_2$ in clonal cell lines

In 1973 Powis showed that noradrenaline is transported into primary cultured embryonic tracheal cells. The K$_m$ was similar to that of uptake$_2$. However, the mechanism does not seem to be identical with uptake$_2$, since the IC$_{50}$ of normetanephrine exceeded that reported for uptake$_2$ by a factor of 50 (Iversen, 1965).

The clonal human kidney cancer cell line Caki-1 is the first cell line which has been reported to possess uptake$_2$. The conclusion that Caki-1 cells express uptake$_2$ is confirmed by several experimental findings. (1) Noradrenaline transport into Caki-1 cells was saturable, the K_m being 450 μmol/l. (2) Inhibitors of uptake$_2$ such as 1 μmol/l corticosterone and 100 μmol/l O-methylisoprenaline largely inhibited noradrenaline uptake in Caki-1 cells. On the contrary, inhibitors of uptake$_1$ such as 1 μmol/l desipramine and 10 μmol/l cocaine did not reduce it. (3) Depolarization of Caki-1 cells by the elevation of extracellular potassium inhibited noradrenaline uptake. (4) There was a highly significant correlation between the IC_{50}'s of various compounds for the inhibition of noradrenaline uptake in Caki-1 cells and rabbit aorta known to possess uptake$_2$ (Schömig and Schönfeld, 1990).

Interestingly enough, it was not possible to demonstrate uptake$_2$ in cell lines which stem from tissues known to possess uptake$_2$ such as A10 rat vascular smooth muscle cells and H9c2 rat myoblasts (Schömig et al., 1990). Most recently a second cell line with uptake$_2$ has been identified, namely the human FL amnion cell line (de la Lande et al., 1991).

Caki-1 cells as a metabolizing system

Trendelenburg (1984) pointed out that the corticosteron-sensitive extraneuronal inactivation of noradrenaline is brought about by a two step metabolizing system. In a first step, noradrenaline is taken up from the extracellular space by uptake$_2$ and in a second step the intracellular enzymes catechol-O-methyltransferase (COMT) and monoamineoxidase (MAO) metabolize the amine. The question was addressed, whether Caki-1 cells may serve as an experimental system for such an extraneuronal metabolizing system.

Incubation of Caki-1 cells for 60 minutes with 100 nmol/l ^{3}H-noradrenaline (COMT and MAO intact) resulted in the formation of deaminated and O-methylated metabolites of noradrenaline (Table 1). In Caki-1 cells, COMT is more active than MAO. Inhibition of COMT by 10 μmol/l U-0521 increased the intracellular steady-state noradrenaline level by a factor of 7, whereas inhibition of MAO by 10 μmol/l pargyline failed to influence intracellular noradrenaline levels at all (Table 1). When intracellular metabolizing enzymes were intact, uptake$_2$ was not able to produce a tissue/medium ratio for noradrenaline which exceeded 0.25. Thus, the rate limiting step of this extraneuronal metabolizing system seems to be transmembrane transport of noradrenaline and not subsequent intracellular metabolism.

The metabolizing capacity of Caki-1 cells incubated with 100 nmol/l noradrenaline amounted to 60 fmol/(min · mg cell protein). The comparison on the basis of intracellular volume with the rate constants for extraneuronal metabolism published by Grohmann (1987) reveals that Caki-1 cells metabolize noradrenaline only by a factor of about three less effectively than the isolated perfused rat heart.

Table 1. Steady-state accumulation and metabolization of ^{3}H-noradrenaline by Caki-1 cells

	NA T/M-ratio	OMDA	NMN	DOPEG	DOMA
			pmol/(mg protein .60 min)		
Control	0.21 ± 0.01	1.40 ± 0.12	2.00 ± 0.10	<0.05	<0.05
U-0521	1.28 ± 0.05	1.01 ± 0.15	0.21 ± 0.04	0.24 ± 0.03	<0.05
Pargyline	0.10 ± 0.01	1.12 ± 0.17	1.87 ± 0.17	<0.05	<0.05
Corticosterone	0.18 ± 0.01	0.35 ± 0.05	0.40 ± 0.07	<0.05	<0.05

Caki-1 cells were incubated with 100 nmol/l ^{3}H-noradrenaline for 60 min at 37°C in the absence (control) or presence of a COMT inhibitor (U-0521; 10 μmol/l), a MAO inhibitor (pargyline; 10 μmol/l), or an uptake$_2$ inhibitor (corticosterone; 10 μmol/l). The inhibitors were present during 20 min of preincubation and during the incubation period. Intracellular ^{3}H-noradrenaline and its ^{3}-H-metabolites as well as ^{3}H-metabolites in the incubation medium were measured at the end of the incubation period. The tissue-medium ratio for ^{3}H-noradrenaline (NA T/M-ratio) was calculated from the amount of intracellular ^{3}H-noradrenaline and from the water space of Caki-1 cells which has been determined in separate experiments. In addition, the Table gives the amount of metabolite formation which has been calculated as the sum of intracellular ^{3}H-metabolite and extracellular ^{3}H-metabolite at the end of the incubation period (*OMDA* 3-methoxy-4-hydroxyphenylglycol, *NMN* normetanephrine, *DOPEG* 3,4-dihydroxyphenylglycol, *DOMA* 3,4-dihydroxymandelic acid). Given are means with SEM's (n = 4)

Similarities between uptake$_2$ and the renal transport system for organic cations

The characterization of uptake$_2$ in Caki-1 cells disclosed a striking similarity between uptake$_2$ and the renal transport mechanism for organic cations (RTOC). RTOC exists in renal tubular cells and is involved in the transport of weak bases from the blood stream into the renal tubular lumen. A variety of compounds have been identified as substrates of RTOC. This list includes endogenous compounds such as dopamine, adrenaline, noradrenaline, acetylcholine, choline, creatinine, histamine, thiamine, and 5-hydroxytryptamine as well as exogenous compounds such as amiloride, isoprenaline, quinidine, procainamide, and cimetidine (Rennick, 1981).

Originally, the hypothesis of a close relationship between uptake$_2$ and RTOC was based (1) on the finding that uptake$_2$ exists in the renal Caki-1 cell line and (2) on similarities of the substrate spectra. Adrenaline, noradrenaline, dopamine and isoprenaline are substrates of both uptake$_2$ (Trendelenburg, 1988) and RTOC (Rennick, 1981). The close relation between the two transport systems was further substantiated by experimental findings. Firstly, the organic cations cimetidine, quinidine and procainamide — which are substrates of RTOC — inhibit RTOC and uptake$_2$ with very similar potencies (Fig. 1). Secondly, cimetidine was shown to be transported via uptake$_2$ in the isolated perfused rat heart (Schömig and Schönfeld, 1990).

The surprising analogy between the characteristics of uptake$_2$ and RTOC was further investigated. As experimental model for RTOC the

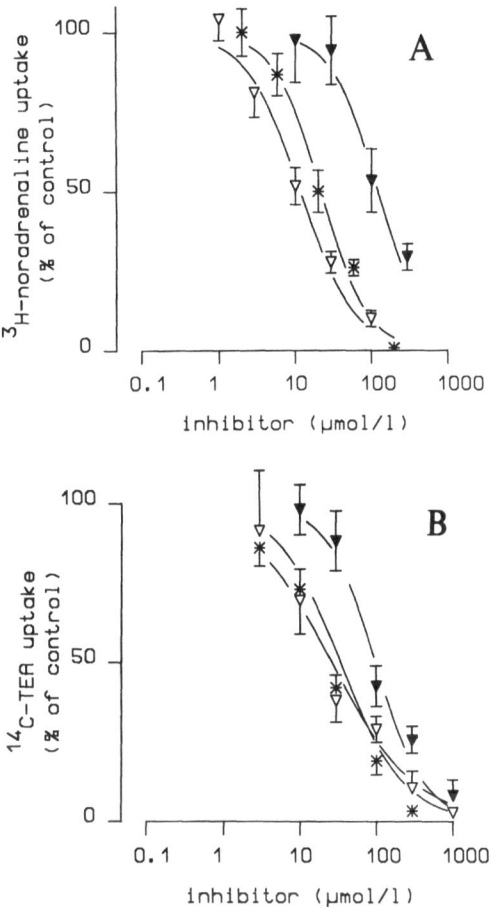

Fig. 1. Inhibition of uptake$_2$ in Caki-1 cells and of RTOC in LLC-PK$_1$ cells **A** Initial rates of ^{3}H-noradrenaline transport into Caki-1 cells were determined in the presence of various concentrations of quinidine (*), cimetidine ($\triangledown$), or procainamide ($\blacktriangledown$). Shown are means ±SEM (n = 5-6). Data are taken from Schömig and Schönfeld (1990) **B** Initial rates of ^{14}C-TEA transport into LLC-PK$_1$ cells were determined in the presence of various concentrations of each inhibitor. LLC-PK$_1$ cells were incubated for 120 s at 37°C with 4 µmol/l ^{14}C-TEA either in the absence (control) or in the presence of quinidine (*), cimetidine ($\triangledown$), or procainamide ($\blacktriangledown$). Shown are means ±SEM (n = 4-5)

transport of the organic cation ^{14}C-tetraethylammonium (^{14}C-TEA) into the renal LLC-PK$_1$ cell line was used. LLC-PK$_1$ cells are a well established model for RTOC (McKinney et al., 1988; Fouda et al., 1990). Confluent monolayers of LLC-PK$_1$ cells were incubated for 120 s at 37°C with ^{14}C-tetraethylammonium in order to measure initial rates of organic cation transport through the apical plasma membrane of LLC-PK$_1$ cells. The inhibitory potencies of various steroids were determined. Steroids were chosen, since corticosterone and 17β-estradiol are the most potent inhibitors of uptake$_2$ known so far. An interaction between steroids and the RTOC has not yet been described.

Table 2. Inhibition by various steroids of the renal transport mechanism for organic cations (RTOC) and uptake$_2$

	RTOC K_i (μmol/l)	Uptake$_2$ K_i (μmol/l)
Progesterone	0.15	1.5
Corticosterone	0.26	0.14
Testosterone	0.61	1.5
17β-Estradiol	2.0	0.36
Hydrocortisone	3.2	3.2
Aldosterone	25	4.2

Initial rates of ^{14}C-tetraethylammonium transport into LLC-PK$_1$ cells (RTOC) or initial rates of ^{3}H-noradrenaline transport into Caki-1 cells (uptake$_2$) were measured in the presence of various concentrations of each steroid. IC$_{50}$'s were calculated from inhibition data by non-linear least square regression analysis, and transformed to corresponding K$_i$'s. Given are means of 4 experiments. There was no significant correlation between the logarithms of the K$_i$'s for the inhibition of RTOC and for the inhibition of uptake$_2$ ($r = 0.5567$; $n = 6$; n.s.)

In fact, various steroids were found to inhibit RTOC. Progesterone, corticosterone, testosterone and testosterone acetate potently inhibited ^{14}C-TEA uptake into LLC-PK$_1$ cells (Table 2). Progesterone is the most potent inhibitor of RTOC known so far. Hydrocortisone and dexamethasone were less potent inhibitors and the K$_i$ of aldosterone was above 10 μmol/l (Table 2).

In Table 2, the K$_i$'s of various steroids for the inhibition of RTOC in LLC-PK$_1$ cells are compared with the corresponding K$_i$'s for the inhibition of uptake$_2$ in Caki-1 cells. Although, some steroids inhibited both transport systems with rather high potencies, there is no significant correlation between the potencies for the inhibition of uptake$_2$ and RTOC. The missing correlation indicates that RTOC and uptake$_2$ are not identical. However, the finding that steroids potently inhibit both uptake$_2$ and RTOC further supports the view of a very close relationship between these transport systems.

Acknowledgements

We thank E. Fekete and M. Hoffmann for skilful technical assistance. This work was supported by the Deutsche Forschungsgemeinschaft (SFB176).

References

American Type Culture Collection (ATCC) (1988) Catalogue of cell lines and hybridomas, 6th edn. Rockville, Maryland

Bryan LJ (1990) How many different uptake processes exist for catecholamines? J Auton Pharmacol 10: 3–4

Coppeneur D, Lingen B, Bönisch H (1991) Expression of the neuronal noradrenaline transporter in xenopus laevis oocytes. Naunyn-Schmiedebergs Arch Pharmacol 343 [Suppl]: R91

de la Lande IS, Marino V, Lavranos T, Kennedy JA, Parker DAS, Seamark RF (1991) Distribution of extraneuronal uptake₁ in reproductive tissues: studies on cells in culture (this volume)

Fouda A-K, Fauth C, Roch-Ramel F (1990) Transport of organic cations by kidney epithelial cell line LLC-PK₁. J Pharmacol Exp Ther 252: 286–292

Graefe K-H, Bönisch H (1988) The transport of amines across the axonal membranes of noradrenergic and dopaminergic neurones. In: Trendelenburg U, Weiner N (eds) Catecholamines I. Springer, Berlin Heidelberg New York Tokyo, pp 193–245 (Handb Exp Pharmacol 90)

Graefe K-H, Stefano FJE, Langer SZ (1973) Preferential metabolism of ³H-(-)-norepinephrine through the deaminated glycol in the rat vas deferens. Biochem Pharmacol 22: 1147–1160

Grohmann M (1987) The activity of the neuronal and extraneuronal catecholamine-metabolizing enzymes of the perfused rat heart. Naunyn-Schmiedebergs Arch Pharmacol 336: 139–147

Iversen LL (1965) The uptake of adrenaline by the rat isolated heart. Br J Pharmacol 24: 387–394

Kennedy JA, de la Lande IS (1987) Characteristics of the cocaine-sensitive accumulation and O-methylation of ³H-(-)-noradrenaline by rabbit endometrium. Naunyn-Schmiedebergs Arch Pharmacol 336: 148–154

McKinney TD, DeLeon C, Speeg KV (1988) Organic cation uptake by a cultured renal epithelium. J Cell Physiol 137: 513–520

Pacholczyk T, Blakely RD, Amara SG (1991) Expression cloning of a cocaine- and antidepressant-sensitive human noradrenaline transporter. Nature 350: 350–354

Parker DAS, de la Lande IS, Proctor C, Marino V, Lam NX, Parker I (1987) Cocaine-sensitive O-methylation of noradrenaline in dental pulp of the rabbit: comparison with the rabbit ear artery. Naunyn-Schmiedebergs Arch Pharmacol 335: 32–39

Powis G (1973) The accumulation and metabolism of (-)-noradrenaline by cells in culture. Br J Pharmacol 47: 568–575

Rennick BR (1981) Renal tubule transport of organic cations. Am J Physiol 240: F83–F89

Schömig E, Schönfeld C-L (1990) Extraneuronal noradrenaline transport (uptake₂) in a human cell line (Caki-1 cells). Naunyn-Schmiedebergs Arch Pharmacol 341: 404–410

Schömig E, Babin-Ebell J, Schönfeld C-L, Russ H, Trendelenburg U (1990) Human Caki-1 cells are the first model for extraneuronal transport of noradrenaline (uptake₂) which is based on a clonal cell line. J Neural Transm [Suppl] 32: 437–440

Trendelenburg U (1984) Metabolizing systems. In: Fleming WW, Langer SZ, Graefe KH, Weiner N (eds) Neuronal and extraneuronal events in autonomic pharmacology. Raven Press, New York, pp 93–109

Trendelenburg U (1988) The extraneuronal uptake and metabolism of catecholamines. In: Trendelenburg U, Weiner N (eds) Catecholamines I. Springer, Berlin Heidelberg New York Tokyo, pp 279–319 (Handb Exp Pharmacol 90)

Authors' address: Dr. E. Schömig, Department of Pharmacology, University of Würzburg, Versbacher Strasse 9, D-W-8700 Würzburg, Federal Republic of Germany

J Neural Transm (1991) [Suppl] 34: 51–59

The influence of the estrous cycle on the accumulation of ^{3}H-noradrenaline in rat uterus and on efflux of radioactivity from the uterus

I. Negrón and **S. Pluchino**

Department of Pharmacology, School of Medicine, Universidad Central de Venezuela,
Caracas, Venezuela

Summary. Groups of female rats were used throughout the 4-days estrous cycle and the accumulation of noradrenaline and its efflux were studied after incubation with the labelled amine. Comparisons were made between uterine horns corresponding to each one of the 4 stages of the cycle, i.e., diestrus, proestrus, metestrus and estrus.

Accumulation in diestrus reached the highest value, which was 1.64-fold higher than in estrus.

An estimate of the original distribution of the amine into the extraneuronal and neuroanl compartments of the tissue was obtained by compartmental analysis of the efflux curves.

Size and half times of neuronal and extraneuronal compartments showed some relationship with the stage of the cycle.

Introduction

Since 1973, the involvement of noradrenaline (NA) and dopamine has been attested in neuroendocrine regulation, particularly in the control of gonadotropin secretion (Kalra and McCann, 1973).

On the other hand, changes have been demonstrated in NA content of some brain regions throughout the 4-days estrous cycle of female rats (Crawley et al., 1978). The reduction of NA content in the medial preoptic and paraventricular nuclei during proestrus-estrus may indicate enhanced noradrenergic neurotransmission while a decreased noradrenergic neuro-transmission might occurr during metestrus-diestrus.

There is widespread agreement that gonadal secretions reduce the turn-over of cerebral NA, which is increased following castration (Munaro, 1977). The decreases in turnover produced by estrogen therapy are evidently reversed in a number of hypothalamic areas by progesterone. It is likely that the regional changes in catecholamine levels during the estrous cycle are due to fluctuations in circulating gonadal steroids, particularly estrogen. The determination of plasma progesterone and estradiol-17β throughout the 4-days estrous cycle of the rat showed an elevated level of estradiol on the day of proestrus, while two peaks of progesterone occurred in plasma during the estrous cycle, a first smaller peak representing the short functional life of corpora lutea and a preovulatory higher peak (Butcher et al., 1974).

Gonadal steroid treatment results in the alteration of a number of neural enzyme activities. The estrogen-dependent decrease of type A monoamine oxidase (MAO), is due to an increased rate of degradation rather than a decreased rate of synthesis (Luine and McEwen, 1977). Direct steroid effects on membrane transport are seen at micromolar estradiol concentrations, which are able to block catecholamine uptake$_2$ in isolated rat heart (Iversen and Salt, 1970).

Almost all of the central areas exhibiting changes in catecholamine content over the estrous cycle are target tissues for estrogen. An interesting target for studying nongenomic effects of steroid hormones on presynaptic events would be the uterus. Adrenergic nerves of the rat uterus are associated with smooth muscle, blood vessels, and glandular elements; moreover, it has been demonstrated that isolated rat uterine horns are able to take up ^{3}H-NA, store, and protect it in some subcellular structure (Fernandez-Pardal et al., 1981).

The present study was undertaken in order to explore whether the influence of sex hormones is able to modify the accumulation and the spontaneous efflux of NA. We determined efflux curves after an initial incubation with radioactively labelled NA; we also evaluated the distribution of the amine from efflux curves (Bönisch et al., 1974).

Material and methods

Virgin female adult Wistar rats (200–250 g b.w.) were kept in a temperature controlled room (20°C) with 10 h with white light and 14 hours of darkness. Vaginal smears were taken daily to monitor the estrous cycle. Only rats which exhibited at least 4 regular cycles lasting 4 days were used. For the compartmental analysis, tissues were obtained from rats which exhibited one of the four phases of the estrous cycle: proestrus (P), estrus (E), metestrus (M), and diestrus (D). In all cases the animals were stunned by a blow on the neck and their uterine horns removed and placed into 3 ml of solution kept at 37°C and bubbled with 95% O_2 and 5% CO_2. The composition of the solution (mmol/l) was as follows: NaCl, 140; KCl, 5; $CaCl_2$, 2.5; Tris-HCl, pH 7.8, 10; glucose 1.1; Na_2-EDTA, 0.04; ascorbic acid, 0.11. The final pH of the solution was 7.5 at 37°C. After preincubation for 15 min, the tissue was transferred into 3.0 ml of medium containing (−)-^{3}H-NA (25×10^{-7} mol/l) and incubated for 30 min at 37°C. Then, the tissue was washed for 180 min with amine-free medium by transferring it into new tubes each containing 3 ml of medium. The washing was accomplished according to the following scheme: 5 washings of 2 min each, 14 of 5 min, and 10 of 10 min. The efflux of radioactivity was measured by collecting 1-ml samples for each tube. The samples were pooled and acidified to pH 2 by the addition of 2 mol/l HCl.

At the end of the experiment, the tissues were blotted dry, weighed and homogenized in 3 ml of ice-cold 0.4 mol/l perchloric acid containing 1 mg/ml of EDTA and 1.25 mg/ml of Na_2SO_3. After 30 min of cold storage, tissue samples were centrifuged for 10 min at 10.000 xg.

The radioactivity in tissue and medium was determined by scintillation counting (for details, see Graefe et al., 1973).

After preincubation, in order to block MAO and COMT, uterine horns were exposed to pargyline and to U-0521 (for details see Henseling et al., 1976).

The wash-out curves obtained in our experiments were multiphasic exponential and can be considered as originating from several single exponential processes. The compartmental analysis was based on the description made by Henseling et al. (1976). In our case, the compartment with the longest half time was called compartment III; the compartments with shorter half times were called as compartment II and I, respectively. We calculated the "bound fraction" from the regression line describing compartment III, obtaining not only the size of the compartment but also the amount of radioactivity that was lost from this compartment up to the end of the experiment. The difference between these 2 values equals the amount of radioactivity left in compartment III at the end of the experiment. The "bound fraction" then equals the difference between "total radioactivity recovered from uterine horns at the end of the experiment" and "radioactivity attributable to compartment III."

In all efflux experiments the total tissue accumulation at zero time was estimated by adding the radioactivity recovered from the efflux samples to that recovered from the tissue. The analysis of efflux curves was based on rates, i.e., on $pmol \cdot g^{-1} \cdot min^{-1}$.

Results were expressed as means $\pm$ S.E.; n indicates the number of observations. The significance of differences was analysed with Student's t-test or with paired t-test.

The agents used in this study were the following: $(-)$-noradrenaline-7-^{3}H (NEN Chemicals; 6.4 Ci/mmol); pargyline hydrochloride (Abbott); 3',4'-dihydroxy-2-methylpropiophenone (U-0521, Upjohn).

Results

Accumulation of radioactivity in rat uterine horns during incubation with ^{3}H-NA

Accumulation at zero time was determined in each one of the estrous stages (5 determinations per stage).

In all tissues analysed, the highest accumalation was found during diestrus and proestrus (6421.7 $\pm$ 224.1 and 5924.3 $\pm$ 240.4 $pmol \cdot g^{-1}$, respectively; mean $\pm$ S.E., P > 0.5) (Fig. 1). In contrast, the uterus showed the lowest accumulation during metestrus and estrus (4263.3 $\pm$ 215.2 and 3949.3 $\pm$ 223.5 $pmol \cdot g^{-1}$, respectively; mean $\pm$ S.E., P > 0.5). Accumulation in diestrus was 1.63-fold higher than in estrus (P < 0.001) and 1,5-fold higher than in metestrus (P < 0.001); similarly, highly significant differences (P < 0.001) were found when proestrus was compared either with metestrus or with estrus.

Efflux of radioactivity after incubation with NA

Uterine horns previously treated with pargyline were incubated with ^{3}H-NA and U-0521 (for details, see Methods); combined treatment with both agents reduces metabolite formation by about 95% (Henseling et al., 1976). After the initial incubation with the labelled amine, the tissues were washed

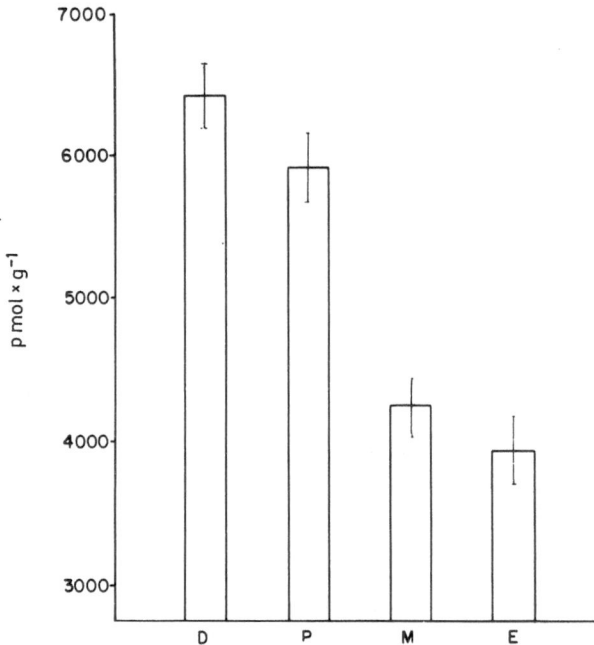

Fig. 1. Accumulation of radioactivity in uterine horns incubated with labelled noradrenaline. Height of columns: accumulation of radioactivity (in pmol $\times$ g^{-1}). Diestrus (D), proestrus (P), metestrus (M) and estrus (E) were the estrous cycle stages studied. Shown are means $\pm$ S.E. of 5 observations each

for 180 min. For each one of the stages of the estrous cycle, our efflux curves were multiphasic exponential; during the late phase of wash-out (from the 80th min of wash-out onward) the curves showed a linear decline which indicated that, at this time, the efflux of radioactivity originated predominantly from a single compartment. The rates of efflux of radioactivity declined steeply during early wash-out, while during the second half of the experiment the rates of efflux decreased.

The stage of the estrous cycle influenced the rates of efflux of radioactivity: these were higher in diestrus than in estrus (min 10 and 30, $P < 0.001$; min 50, $P < 0.005$; min 70, $P < 0.01$; min 120, $P < 0.02$; min 180, $P < 0.005$) (Fig. 2). The rates of efflux observed in proestrus did not differ from those described for diestrus; similarly, the efflux curve for estrus was virtually identical with that obtained for metestrus. Finally, the rates of efflux observed in proestrus were higher than those determined in estrus.

The compartmental analysis of the wash-out curves for all stages of the estrous cycle shows that most of the radioactivity leaving the uterine horns was attributed to compartment III which is characterized by half times longer than 3 h (Tables 1 and 2): this compartment mainly corresponds to intra-neuronal accumulation of the labelled amine. This compartment reached its maximum size in diestrus, in which the size of compartment III was greater than in estrus and in metestrus (Table 1). On the other hand, the size of compartment III did not differ when diestrus and proestrus were com-

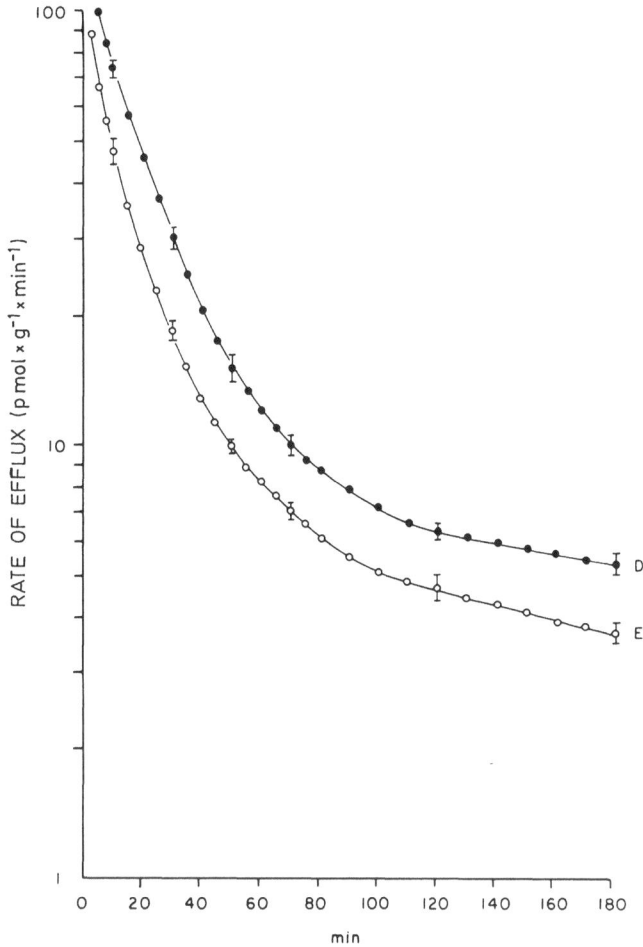

Fig. 2. Efflux of total radioactivity during wash-out of uterine horns after an initial incubation with ³H-noradrenaline. *Ordinate*: rate of efflux (in pmol . g⁻¹ . min⁻¹; log scale); *abscissa*: time (in min) after beginning of wash-out with amine-free solution. Shown are means (± S.E.) for 5 tissues in diestrus (D) and 5 tissues in estrus (E)

pared; in a similar way, no differences were observed between estrus and metestrus.

Compartment II seems to correspond to the extraneuronal accumulation of the amine; the relationship between the sizes of different stages followed the same pattern as for compartment III. Table 1 also shows that for the size of compartment I no differences were observed when comparisons were made between the various stages of the estrous cycle.

As far as half times are concerned, compartment II (10.9 to 12.9 min) and compartment I (1.4 to 1.5 min) did not show differences when values for different stages were compared, although the half time for the efflux from compartment II was longer than that for compartment I.

Uterine horns in diestrus had a considerable "bound fraction" which was significantly greater than that for metestrus or estrus (Table 1); this portion

Table 1. Compartmental analysis of efflux curves obtained during wash-out of uterine horns previously exposed to ³H-NA

Stage	n	III* (pmol·g⁻¹)	P	II** (pmol·g⁻¹)	P	I*** (pmol·g⁻¹)	P	B.F.**** (pmol·g⁻¹)	P	Accumulation (pmol·g⁻¹)
M(a)	5	2191.8 ± 116.9	<0.001	1341.6 ± 75.5	<0.001	439.1 ± 18.1	N.S.	291.1 ± 5.4	<0.001	4263.6 ± 215.2
D(b)	5	3023.2 ± 73.5	<0.001	2017.3 ± 47.5	<0.001	466.9 ± 20.4	N.S.	914.3 ± 96.6	<0.001	6421.7 ± 224.1
E(c)	5	2105.4 ± 152.4	<0.001	1238.3 ± 32.0	<0.001	367.2 ± 46.9	N.S.	238.4 ± 25.1	<0.001	3949.3 ± 223.5
P(d)	5	2876.1 ± 131.9	<0.001	1868.7 ± 55.4	<0.001	467.1 ± 30.1	N.S.	712.4 ± 35.1	<0.001	5924.3 ± 240.4

Shown are means ±S.E. for n experiments. *M* metestrus; *D* diestrus; *E* estrus; *P* proestrus. *B.F.* bound fraction. *N.S.* not significant.

* a-c: NS; a-d: P < 0.005; b-d: NS
** a-c: NS; a-d: P < 0.001; b-d: NS
*** a-c: NS; a-d, b-d: NS
**** a-c:NS; a-d: P < 0.001; b-d: NS

Table 2. Half times of compartments

Stage	n	III* (min)	P	II* (min)	P	I*** (min)	P
M(a)	5	195.3 ± 2.2		10.9 ± 1.1		1.5 ± 0.09	
D(b)	5	226.8 ± 7.7	<0.005	12.9 ± 0.8	NS	1.4 ± 0.06	NS
E(c)	5	190.9 ± 10.0	<0.025	11.8 ± 1.2	NS	1.4 ± 0.05	NS
P(d)	5	221.9 ± 5.1	<0.025	11.9 ± 0.9	NS	1.5 ± 0.07	NS

Shown are mean ±S.E. for n experiments. For further abbreviations, see Table 1.
* a-c: NS: a-d: p < 0.005; b-d: NS
** a-c, a-d, b-d: NS
*** a-c, a-d, b-d: NS

of the accumulated radioactivity does not contribute to the efflux and, therefore, it does not affect the compartmental analysis (Henseling et al., 1976).

Discussion

Catecholamines are removed from the synaptic gap by two main processes; uptake into neuronal tissue (uptake$_1$) and uptake into extraneuronal tissue (uptake$_2$). It is possible that variations in the uptake of the catecholamines could occur in the different stages of the estrous cycle, since levels of gonadal hormones fluctuate during the cycle (Butcher et al., 1974). In this study, the changes of the accumulation within the four stages of the cycle revealed significant differences in the uptake of ^{3}H-NA. The pattern of accumulation for each stage indicated that D group reached higher accumulation compared with E group; similar relationships were observed when the "bound fraction" and the size of compartments II and III were compared. The magnitude of the difference of accumulation was mainly due to the sum of the radioactivity contained in "bound fraction" and compartment III. From these experiments it can be concluded that the difference in the accumulation of radioactivity between group D and E was mainly due to the neuronal accumulation of radioactivity.

When uterine horns were exposed to high concentrations of estradiol (estrus stage), the incubation with NA mainly resulted in a decrease in the size of compartment III and of the "bound fraction", which support the view that compartment III and "bound fraction" represent an accumulation of radioactivity in the neurone (axoplasmic and/or vesicular). While Ghraf et al. (1983) described a stereospecific competition of NA uptake into synaptosomes from rat central structures in the presence of high concentrations $(10^{-6}-10^{-5}\text{mol/l})$ of estradiol-17β, our results demonstrated this mechanism under extreme physiological conditions.

Carrier-mediated transport and passive diffusion contribute to the spontaneous neuronal efflux (Carlsson and Waldeck, 1968; Paton, 1981); therefore, it can be concluded that the spontaneous neuronal efflux is in part carrier-mediated. An influence of estradiol-17β on the rat uterine Na^+/K^+ ATPase activity has been demonstrated; 50 μg estradiol-17β decrease ATPase activity by approximately 40% (Karmakar, 1969). These findings are in agreement with our results. In effect, the uterine tissues in E (i.e., on exposure to high plasma concentrations of estradiol-17β) reached a lower accumulation than tissues in D (this stage is far away from the estrogen peak); moreover, slower rates of efflux and a shorter half time characterized the estrus stage.

There was efflux of amine from the uterine horns with a half time of <2 min; quantitatively, this efflux corresponded to the amine distributed into the extracellular space. It is reasonable to conclude that compartment II (with a half time of about 10 min) represented the extraneuronal store of the amine. As expected, the size of compartment II was smaller than that of compartment III in each one of the stages, and it was sensitive to the exposure to high concentrations of estradiol-17β.

References

Bönisch H, Uhlig W, Trendelenburg U (1974) Analysis of the compartments involved in the extraneuronal storage and metabolism of isoprenaline in the perfused heart. Naunyn-Schmiedebergs Arch Pharmacol 283: 223–244

Butcher Rl, Collins WE, Fugo NW (1974) Plasma concentration of LH, FSH, prolactin, progesterone and estradiol-17β throughout the 4-day estrous cycle of the rat. Endocrinology 94: 1704–1708

Carlsson A, Waldeck B (1968) Different mechanism of drug-induced release of noradrenaline and its congeners α-methylnoradrenaline and metaraminol. Eur J Pharmacol 3: 165–168

Crawley WR, O'Donohue TL, Jacobowitz DM (1978) Changes in catecholamine content in discrete brain nuclei during the estrous cycle of the rat. Brain Res 147: 315–326

Fernandez-Pardal J, Gimeno M, Gimeno A (1981) Metabolism of ^{3}H-noradrenaline by the isolated rat uterus. Can J Physiol Pharmacol 59: 1245–1249

Ghraf R, Michel M, Hiemke C, Knuppen R (1983) Competition by monophenolic estrogens and catecholestrogens for high-affinity uptake of ^{3}H-norepinephrine into synaptosomes from rat cerebral cortex and hypothalamus. Brain Res 277: 163–168

Graefe KH, Stefano F, Langer SZ (1973) Preferential metabolism of ^{3}H-norepinephrine through the deaminated glycol in the rat vas deferens. Biochem Pharmacol 22: 1147–1160

Henseling M, Eckert E, Trendelenburg U (1976) The distribution of ^{3}H-noradrenaline in rabbit aortic strips after inhibition of the noradrenaline-metabolizing enzymes. Naunyn-Schmiedebergs Arch Pharmacol 292: 205–217

Iversen LL, Salt PJ (1970) Inhibition of catecholamine uptake$_2$ by steroids in the isolated rat heart. Br J Pharmacol 40: 528–530

Kalra PS, McCann SM (1973) Involvement of catecholamines in feedback mechanisms. Prog Brain Res 39: 185–198

Karmakar PK (1969) The influence of oestradiol-17β on the rat uterine Na^+, K^+-Mg^{++} activated adenosine triphosphatase activity. Experientia 25/3: 319–320

Luine VN, McEwen BS (1977) Effect of oestradiol on turnover of type A monoamine oxidase in brain. J Neurochem 28: 1221–1227

Munaro NI (1977) The effect of ovarian steroids on hypothalamic norepinephrine neuronal activity. Acta Endocrinol 86: 235–242

Paton DM (1981) Effect of cocaine on the efflux of noradrenaline, octopamine and metaraminol in rabbit atria. IRCS Med Sci 9: 128

Authors' address: Dr. S. Pluchino, Department of Pharmacology, U.C.V. Caracas, Apartado 19.253 — Qta. Crespo, Caracas 1010-A, Venezuela

J Neural Transm (1991) [Suppl] 34: 61–67

PC12 cells as a window for the differentiation of neural crest into adrenergic nerve ending and adrenal medulla

M. B. H. Youdim

Technion — The Bruce Rappaport Faculty of Medicine, Department of Pharmacology,
Haifa, Israel

Summary. Studies on PC12 and isolated adrenal chromaffin cells have revealed that PC12 cells have a closer identity to the adrenergic nerve ending than do the chromaffin cells. This is revealed by the presence of monoamine oxidase (MAO) A and tyramine-released pool of catecholamines in PC12, resembling that in adrenergic neurones, and their absence in adrenal chromaffin cells. Indeed, chromaffin cells possess primarilly MAO-B activity. Like the observations on adrenergic neurones, non-selective and selective MAO-A inhibitors potentiate the catecholamine-releasing property of tyramine in PC12 cells. This property has clearly been demonstrated to be associated with selective inhibition of MAO-A and not MAO-B. The fact that MAO-A and MAO-B are different proteins and under separate gene product control suggests that their regulation may be highly differentiated. Indeed, it has been shown that while steroids such as progesterone and hydrocortisone induce and estrogen diminishes MAO-A activity in PC12 cells, no such regulatory mechanism has been identified for MAO-B activity in chromaffin cells. In the final analysis the inter-relationship between MAO-A activity and the presence of tyramine-releasable pool of catecholamines in adrenergic neurons and PC12 cells may have a genetic basis and could be important in illuminating the differentiation of neural crest into adrenergic neurones and adrenal medulla on the one hand and chromaffin cells to PC12 cells on the other.

Introduction

PC12 (phaeochromocytoma) cells, derived from adrenal medulla, like adrenergic nerve ending and adrenal chromaffin cells, possess all the components to synthesize, store, release, take up and catabolize catecholamines (Greene and Rein, 1977a, b; Pollard et al., 1985). Thus, like the chromaffin cell, it has been employed as a model for adrenergic nerve ending (Lee et al., 1977, 1980; Pollard et al., 1985). Such cells have antigenic properties which recognize both the adrenergic neurones and adrenal chromaffin cells, suggesting that it can be a window into the differentiation of neural crest into adrenergic neurone and adrenal medulla (Lee et al., 1977, 1980; Youdim et al., 1986).

Without exception almost all the biochemical and pharmacological studies that have so far been performed on PC12 and chromaffin cells were targeted on the catecholamine synthetic pathway, storage and Ca^{2+}-dependent K^+-induced release. However, until very recently no attention was paid to catecholamine catabolism properties of these cells (Youdim et al., 1984a, b). Further, no explanation was put forward as to why tyramine, an indirectly acting sympathomimetic amine, and monoamine oxidase (MAO) inhibitors do not in vivo induce catecholamine release from adrenal medulla nor alter their metabolism. Since the latter components, together with the presence of monoamine oxidase (MAO) type A, are the hall mark of adrenergic neurone (Jarrott, 1971; Finberg et al., 1981; Finberg and Tenne, 1982; Finberg and Youdim, 1988; Youdim et al., 1988), we investigated these two properties in PC12 and isolated adrenal chromaffin cells in order to establish which, if either, of the two cells better represents the adrenergic neurone.

Monoamine oxidase activity

Mitochondria prepared from homogenized PC12 and chromaffin cells were capable of oxidatively deaminating a wide variety of amines including dopamine, adrenaline, noradrenaline, serotonin, phenylethylamine, tyramine and kynuramine (Youdim et al., 1984a, 1986). The kinetic analysis of amine oxidation by the two cell types revealed that these cells deaminated the monoamines significantly differently to suggest the presence of two different enzymes. Mitochondria from chromaffin cells were highly efficient in oxidation of tyramine, dopamine and phenylethylamine, with apparent Km values similar to those reported for this enzyme in mammalian brain. By contrast, PC12 cell mitochondria monoamine oxidase showed a greater affinity and relative Vmax for adrenaline, noradrenaline, tyramine, dopamine and serotonin. These findings were compatible with PC12 and chromaffin cells having relatively a greater component of MAO type A and MAO type B respectively. Indeed, while PC12 MAO exhibited apparent Km values of $357\,\mu M$ and $400\,\mu M$ respectively for adrenaline and noradrenaline, the apparent Km values for the oxidation of the same amines by chromaffin cells were $1100\,\mu M$ and $1700\,\mu M$. By contrast the MAO B substrates phenylethylamine and benzylamine are avidly oxidized with apparent high affinity Km in the range of $15\,\mu M$. In this respect adrenal chromaffin cell MAO resembles that of the human platelet MAO, which is known to be type B (see Youdim et al., 1988). The low affinity of MAO in chromaffin cells and platelets for adrenaline, noradrenaline and serotonin may be one reason why these cells can accumulate the highest concentrations of catecholamines and serotonin respectively in the body. It is thus possible that in these cells a "wrong" type of MAO for the right reason has been placed as a process for physiological conservation in case of need.

Using the criteria established by Johnston (1968) for differentiation of MAO into its subtypes in response to selective MAO inhibitors clorgyline and l-deprenyl (Knoll and Magyar, 1972), we have now identified that, while PC12 cells contain exclusively MAO type A (Youdim et al., 1986), adrenal chromaffin cells are endowed primarily with MAO-B (Youdim et al., 1984a). These findings are compatible with and complement those regarding the kinetics of biogenic amine oxidation by the two cell types. Thus, in respect to MAO, PC12 cells have a closer affinity to the adrenergic neurones. It is now apparent that this also is the generalized characteristic of human phaeochromocytoma and chromaffin cells, being endowed respectively with the same enzyme forms (Carmichael and Pfeiffer, 1985, 1987; Naoi et al., 1987).

Catecholamine release mechanism in response to tyramine

Like the adrenergic neurone, PC12 and chromaffin cells have well defined voltage- and calcium-dependent K^+-induced catecholamine release mechanisms, as well as uptake systems (Baker and Knight, 1984; Pollard et al., 1985). These processes show great similarities to that of the nerve ending in response to secretagogues and amine uptake blockers. Although in vivo catecholamine release in response to tyramine was examined in cat and rabbit adrenal gland, the release was negligible (Youdim et al., 1986) and MAO inhibitors had no influence (Pollard et al., 1985). Such results are at variance with tyramine's ability to produce catecholamine release from adrenergic neurones where its action is highly potentiated by non-selective and selective MAO-A inhibitors (Finberg et al., 1981; Finberg and Youdim, 1988). We examined, for the first time, the direct effect of tyramine on catecholamine release from PC12 and chromaffin cells as a means to evaluate their uptake and release mechanisms. Both cell types demonstrated tyramine uptake processes, with apparent Km values of 16 μM and 8 μM for PC12 and chromaffin cells, respectively.

The release of endogenous catecholamines and ^{3}[H]-noradrenaline from PC12 and chromaffin cells in response to tyramine (1–1000 μM) and K^+ (50 mM) have revealed that, while both cell types respond to K^+, only PC12 cells show a significant catecholamine release in response to tyramine, which was dose-dependent. The preloading of both cell types with either ^{14}[C]-tyramine or ^{3}[H]-noradrenaline for 24 h gave essentially similar sets of data. Furthermore, the action of tyramine and K^+ in PC12 cells were additive (Youdim et al., 1986).

These studies suggest the presence of two pools of catecholamines in PC12 cells which respond differently to the secretagogues and resemble similar observations in adrenergic nerve endings. Furthermore, similar to adrenergic neurons, non-selective (pargyline and tranylcypramine) (Youdim et al., 1986) and selective MAO-A inhibitors (clorgyline and moclobemide) potentiate tyramine- and K^+-induced catecholamine release from PC12 (Youdim, 1990) but not from chromaffin cells. It is apparent that inhibition of MAO-A in PC12 cells has a significantly greater action on the release of catecholamines as induced by tyramine than by K^+. The lack of tyramine-induced catecholamine release in chromaffin cells cannot be explained by the inability of these cells to take up tyramine. Indeed, these cells, like PC12 cells, do take up significant amounts of tyramine. Thus there may be an absence of a tyramine-releasable pool of catecholamines, or the necessary release receptors. It is also possible that the tyramine taken up is readily metabolized by MAO-B in the chromaffin cells before it can be taken up by the chromaffin granules to initiate the release mechanism. Indeed tyramine initiates a significant release of catecholamines by a displacement phenomenon from isolated purified chromaffin granules.

Differentiation of monoamine oxidases and catecholamine storage compartments

Pharmacological and biochemical studies have revealed two types of MAO designated A and B (Johnston, 1968; Youdim et al., 1988). It is only very recently that these two enzymes have been cloned and their primary structure has been identified to be different (Bach et al., 1988) and under separate gene control (Westlund et al., 1985; Denney and Abell, 1984; Denney and Denney, 1985). No such evidence for two distinct pools of catecholamines has so far been established. The possibility that two different types of granules having variant tyramine uptake or release systems exist in PC12 cells and adrenergic nerve ending, cannot be ruled out. Nevertheless, the co-existence of MAO-A activity and the tyramine releasable pool of catecholamines in PC12 cells and adrenergic neurons may not be coincidental and could have either a genetic basis or be under hormonal control. To investigate this problem, the response of chromaffin cells and PC12 cells in culture to the effects of NGF (nerve growth factor) and several steroids, including progesterone, hydrocortisone, dexamethasone, testosterone and estrogen was examined. Surprisingly, NGF had no net effect on the MAO activities of either of the two cell types in culture. By contrast progesterone and dexamethasone had a profound effect in inducing, while estrogen diminished, the MAO-A activity of PC12 cells (Youdim, 1991) (Fig. 1). These results parallel those recently observed in adrenal medullary capillary endothelial cells, known to also contain only MAO-A (Youdim et al., 1989). By contrast no such hormonal regulatory mechanism could be observed for the MAO-B activity in the adrenal chromaffin cells (Youdim et al., 1989) or any other cell types containing MAO-B activity so far examined. By contrast ganglioside GMI was reported to stimulate expression of extremely low MAO-B activity in PC12 cells devoid of this enzyme (Naoi et al., 1987). No mechanistic explanation so far has been put forward for this observation. The induction and diminution of MAO-A activity in PC12 cells and capillary endothelial cells (Youdim, 1991; Youdim et al., 1989) has been shown to be related to the modulation of enzyme synthesis (Youdim et al., 1989). Thus, it is apparent that hormones mod-

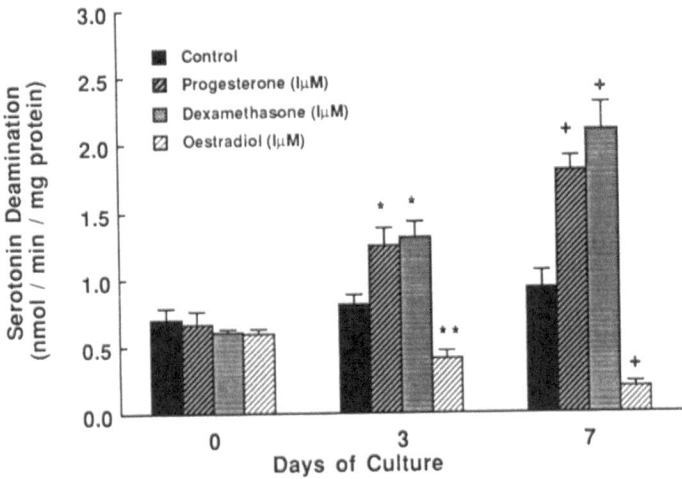

Fig. 1. The effects of steroids progesterone, dexamethasone and oestrogen on the MAO-A activity in PC12 cell cultures (Youdim, 1991)

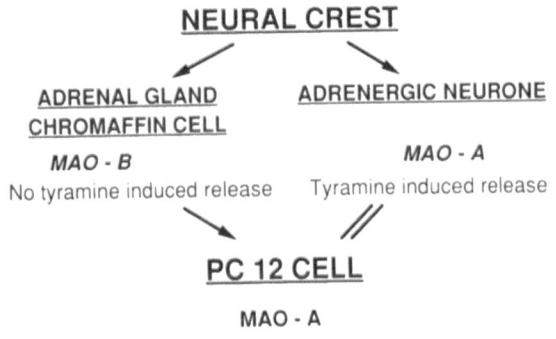

Fig. 2. Schematic pathway for differentiation of monoamine oxidase and tyramine releasable pool of catecholamine in the adrenergic neurone and adrenal medulla as derived from the neural crest (Youdim, 1990)

ulate not the expression or conversion of MAO-A but rather its activity, via enzyme protein synthesis.

The sole expression of MAO-B and MAO-A in the chromaffin and PC12 cells, respectively, is indicative that the genes for the two enzymes must have been derived from neural crest and are present in the chromaffin cell. Therefore, the expression of MAO-B and not MAO-A in the chromaffin cell must be dependent on the presence of genetic promoters for MAO-B and repressor genes specific for MAO-A. During transformation of chromaffin cells to PC12 cells, the expression of such genes may be reversed by oncogenes resulting in supression of MAO-B and expression of MAO-A. One must therefore conclude that similar mechanisms become operative during differentiation of neural crest to adrenergic neurones. Whether the regulation of the tyramine releasable catecholamine pool follows the same path requires further investigation (Fig. 2).

References

Bach AWJ, Lan NC, Johnson DL, Abell CW, Bembenek ME, Kwan SW, Seeburg PH, Shin JC (1988) cDNA cloning of human liver monoamine oxidase A and B: molecular basis of differences in enzymic properties. Proc Natl Acad Sci USA 85: 4934–4938

Baker PF, Knight DE (1984) Calcium control of exocytosis in bovine adrenal medullary cells. TINS 7: 120–126

Carmichael SW, Pfeiffer GL (1985) Histochemical localization of monoamine oxidase type A and B in the adrenal gland. Histochem J 17: 1289–1298

Carmichael SW, Pfeiffer GL (1987) Monoamine oxidase A and B in the human adrenal gland and phaeochromocytoma. Neurochem Int 10: 49–55

Denney RM, Abell CW (1984) The genetics of MAO. In: Tipton K, Dostert P, Strolin-Benedetti M (eds) Monoamine oxidase and disease. Academic Press, London, pp 243–252

Denney RM, Denney CB (1985) An update on the identity crisis of monoamine oxidase: new and old evidence for independence of MAO-A and B. Pharmacol Ther 30: 227–259

Finberg JPM, Tenne M (1982) Relationship between tyramine potentiation and selective inhibition of monoamine oxidase types A and B in the rat vas deferens. Br J Pharmacol 77: 13–21

Finberg JPM, Youdim MBH (1988) Potentiation of tyramine pressor response in conscious rats by reversible inhibitors of monoamine oxidase A. J Neural Transm [Suppl 26]: 11–17

Finberg JPM, Tenne M, Youdim MBH (1981) Tyramine antagonistic properties of AGN 1135, an irreversible inhibitor of monoamine oxidase B. Br J Pharmacol 73: 65–74

Greene LA, Rein G (1977a) Release, storage and uptake of catecholamines by a clonal cell line of nerve growth factor (NGF) response pheochromocytoma cell. Brain Res 129: 247–263

Greene LA, Rein G (1977b) Release of ^{3}H-noradrenaline from a clonal line of pheochromocytoma (PC12) by nicotinic-cholinergic stimulation. Brain Res 138: 521–528

Jarrott B (1971) Occurrence and properties of monoamine oxidase in adrenergic neurons. J Neurochem 18: 7–16

Johnston J (1968) Some observations upon a new inhibitor of monoamine oxidase in brain tissue. Biochem Pharmacol 17: 1285–1297

Knoll J, Magyar K (1972) Some puzzling pharmacological effects of monoamine oxidase inhibitors. Adv Biochem Psychopharmacol 5: 393–408

Lee V, Shelanski ML, Greene LA (1977) Specific neural and adrenal medullary antigens detected by antisera to clonal PC12 and pheochromocytoma cells. Proc Natl Acad Sci USA 74: 5021–5025

Lee VM, Shelanski ML, Greene LA (1980) Characterization of antisera raised against cultured rat sympathetic neurons. Neuroscience 5: 2239–2245

Naoi M, Suzuki H, Takahashi T, Shibahara K, Nagatsu T (1987) Ganglioside GMI causes expression of type B monoamine oxidase in a rat clonal pheochromocytoma cell line PC12. J Neurochem 49: 1600–1605

Pollard HB, Orenberg R, Levine M, Kelner K, Morita K, Levine R, Forsberg E, Brocklehurst KW, Duong L, Elkes RI, Heldman E, Youdim MBH (1985) Hormone secretion by exocytosis with emphasis on information from chromaffin cell system, In: McCormick DB (ed) Vitamins and hormones; advances in research and applications. Academic Press, New York, pp 109–174

Westlund KN, Denney RM, Kochersperger LM, Rose RM, Abell CW (1985) Distinct monoamine oxidase A and B population in primate brain. Science 230: 181–183

Youdim MBH (1990) Monoamine oxidase (MAO)-A but not MAO-B inhibitors potentiated tyramine-induced catecholamine release from PC12 cells. J Neurochem 54: 411–414

Youdim MBH (1991) Regulation of monoamine oxidase A in PC12 cells by steroids. Eur J Pharmacol 192: 201–202

Youdim MBH, Banergee DK, Pollard HB (1984a) Isolated chromaffin cells from adrenal medulla contain primarily monoamine oxidase B. Science 224: 619–621

Youdim MBH, Heldman E, Pollard HB (1984b) Phenotypic changes of MAO activity and catecholamine release characterize the transformation of chromaffin cells to PC12. J Pharm Pharmacol 36: 13–14

Youdim MBH, Heldman E, Pollard HB, Fleming P, McHugh E (1986) Contrasting monoamine oxidase and tyramine induced catecholamine release in PC12 and chromaffin cells. Neuroscience 19: 1311–1318

Youdim MBH, Finberg JPM, Tipton KF (1988) Monoamine oxidase. In: Trendelenburg U, Weiner N (eds) Catecholamine II. Advances in experimental pharmacology, vol 92. Springer, Berlin Heidelberg New York Tokyo, pp 121–197

Youdim MBH, Banergee DK, Offut L, Kelner K, Pollard HB (1989) Steroid regulation of monoamine oxidase activity in the adrenal medulla. FASEB J 3: 1753–1759

Author's address: Prof. M. B. H. Youdim, Department of Pharmacology, The Bruce Rappaport Faculty of Medicine, Bat Galim, Haifa, Israel

J Neural Transm (1991) [Suppl] 34: 69–75

Clinical aspects on presynaptic noradrenaline metabolism

J. Ludwig[1], **T. Halbrügge**[2], and **K.-H. Graefe**[2]

[1] Medical School, and [2] Department of Pharmacology, University of Würzburg,
Würzburg, Federal Republic of Germany

Summary. In healthy subjects, similar absolute increments in plasma noradrenaline (NA) and dihydroxyphenylglycol (DOPEG) were observed in response to upright posture or isoprenaline infusion. Blockade of neuronal uptake by desipramine abolished these plasma DOPEG responses and reduced plasma DOPEG per se. In essential hypertensives we found higher than normal plasma DOPEG levels at any given plasma NA. Evidence is provided that both the desipramine-sensitive and -resistant pool of plasma DOPEG contribute to this hypertensive-normotensive difference.

Introduction

In many clinical investigations plasma noradrenaline levels were taken to reflect sympathetic nervous activity. This is because noradrenaline (NA) in plasma is derived largely from the transmitter released by sympathetic nerves, with only a small fraction originating from the adrenal medulla (Esler et al., 1988). But, there are theoretical objections to this commonly used method, since the overflow of released NA into plasma is influenced by neuronal (uptake$_1$) and extraneuronal (uptake$_2$) uptake, with the most prominent being uptake$_1$. The neuronally recaptured NA is either restored or degraded by monoamine oxidase to mainly form dihydroxyphenylglycol (DOPEG). This metabolite is known to penetrate cell membranes with ease and, hence, to appear in the extracellular fluid at rates corresponding to its formation rates (Trendelenburg et al., 1980). From experimental pharmacology it is evidenced that there are two neuronal sources of DOPEG formation (Fig. 1). One source originates from NA spontaneously leaking out of the transmitter's storage vesicles, the other from NA being recaptured subsequent to its release into the synaptic gap (Graefe and Henseling, 1983; Kopin, 1985). While the latter source of DOPEG formation (uptake source) is sensitive to blockade by uptake$_1$ inhibitors (e.g., desipramine), the former (leakage source) is entirely resistant to blockade of uptake$_1$. Until recently, insufficient attention has been paid to the plasma concentration of DOPEG as an additional index of sympathetic nervous function in humans. Hence, it was of interest to examine whether the above considera-

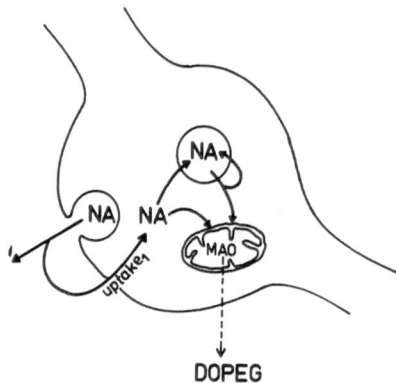

DOPEG

Fig. 1. Schematic representation showing the two sources of neuronal dihydroxyphenylglycol (DOPEG) formation. Noradrenaline (NA), recaptured by neuronal uptake (uptake$_1$) subsequent to exocytotic release, is either transported into storage vesicles or deaminated by monoamine oxidase (MAO) to form DOPEG. This portion of DOPEG formation is susceptible to inhibition by uptake$_1$ blockers (uptake source of DOPEG formation). The second source of DOPEG formation stems from NA spontaneously leaking out of the transmitter's stores and escaping vesicular restorage (leakage source of DOPEG formation). Contrary to the former, the latter source of DOPEG formation is not inhibited by uptake$_1$ blockers

tions based on in vitro experiments also apply, in principle, to the in-vivo condition. The present report is a review of previously published work from our laboratory (Ludwig et al., 1988, 1989, 1991).

Materials and methods

Study 1

After a resting period of 30 min, 12 healthy subjects were exposed to 30 min of quiet sitting as well as 30 min of subsequent quiet standing to enhance sympathetic nervous activity. Forearm venous blood was sampled at the end of each 30-min period via a plastic cannula. Six subjects volunteered to repeat the study after an oral administration of 1.5 mg kg^{-1} desipramine 2-3 h prior to testing. This dose of desipramine is known to produce a high degree of uptake$_1$ inhibiton (Goldstein et al., 1983). The test was now modified in that the resting period was followed only by 30 min of quiet standing (for details, see Ludwig et al., 1988).

Study 2

After 30 min of rest (baseline period), subjects (n = 7) remained supine and were infused i.v. with isoprenaline at constant rates for two consecutive 25-min periods. The isoprenaline infusion rates were 31 to 43 pmol kg^{-1}min^{-1} during the 1st and 80 to 124 pmol kg^{-1}min^{-1} during the 2nd infusion period. At the end of the baseline and the two infusion periods, mixed central venous plasma was sampled from the pulmonary artery via a Swan-Ganz catheter. Five subjects repeated the test 2-3 weeks later after an oral administration of 1.5 mg kg^{-1} desipramine 3 h prior to isoprenaline infusion (for details, see Ludwig et al., 1989).

Study 3

Forty-seven normotensive control subjects and 58 subjects with untreated essential hypertension were studied. The study protocol was identical to that in study 1. The test procedure was repeated in 11 controls and 14 hypertensive subjects after an oral administration of $1.5 \, \text{mg} \, \text{kg}^{-1}$ desipramine as described above (for details, see Ludwig et al., 1991).

Assay procedure

Plasma concentrations of NA and DOPEG were determined by reversed-phase, high-performance liquid chromatography combined with electrochemical detection as described previously (Halbrügge et al., 1988).

Results and discussion

In study 1, orthostasis resulted in similar absolute increases in plasma DOPEG and NA with the magnitude of changes in both parameters being dependent on the degree of orthostasis. For the condition of supine rest, sitting and standing a plot of the mean values of plasma DOPEG (ordinate) against plasma NA revealed a linear relationship between the two parameters. Its slope was about unity, and the regression line intersected the ordinate at a certain value of plasma DOPEG, indicating that even at zero plasma NA (i.e., in the abscence of any sympathetic nervous activity) DOPEG is present in plasma. Thus, two assumptions were made: 1. the increments in plasma DOPEG seen in response to upright posture are derived from the uptake source of DOPEG formation, and 2. the remaining pool of plasma DOPEG at zero plasma NA is derived from the leakage source of DOPEG formation. To substantiate these assumptions some of the subjects were reinvestigated after an oral administration of desipramine to block neuronal NA re-uptake. Under this condition, the plasma concentration of DOPEG fell. Moreover, desipramine abolished the plasma DOPEG response to orthostasis, confirming that increments in plasma DOPEG brought about by orthostasis are presynaptic in origin. Interestingly, the observed mean value of plasma DOPEG in desipramine-treated subjects was virtually identical with the plasma DOPEG value at zero plasma NA obtained from the ordinate intercept of the regression of plasma DOPEG on plasma NA under control conditions. These findings varify the above assumptions, namely that increments in plasma DOPEG observed in response to orthostasis are derived from the uptake source of DOPEG formation, whereas the ordinate intercept of the regression of plasma DOPEG on plasma NA reflects the leakage source of DOPEG formation. It has been proposed that the plasma DOPEG concentration at zero plasma NA stems partly from the central nervous system (Izzo et al., 1985). However, Eisenhofer et al. (1989) argued that if it were so, the plasma DOPEG concentration observed in the presence of desipramine should be lower than that at zero plasma NA. As this was clearly not the case (Ludwig et al., 1988), our results speak in favour of the plasma pool of DOPEG being derived predominantly from peripheral sources of presynaptic DOPEG formation. One further finding of study 1 deserves to be mentioned. It is that desipramine did not change the plasma NA response to orthostasis. We would expect desipramine to enhance the plasma NA response to upright posture, unless desipramine reduces sympathetic traffic. Indeed, recent evidence suggests that desipramine lowers sympathetic activity via a central mechanism of action (Cohen et al., 1990; Szabo and Schultheiss, 1990).

In study 2, two rates of isoprenaline (ISO) were infused i.v. under control conditions and under conditions of desipramine pretreatment, respectively, to study the changes in central venous plasma NA and DOPEG in response to isoprenaline. ISO produced a plasma concentration-dependent increase in plasma NA (Fig. 2A). Desipramine did not alter this response. However, it should be pointed out, that desipramine lowered baseline central venous NA by 47%. This explains why the % increase in plasma NA evoked by ISO was 2.4-fold higher in the presence of desipramine than in its absence. This is in stark contrast to the fact that desipramine failed to enhance the baroreflex-mediated increase in plasma NA due to upright posture (see above). Hence, the ISO-induced increments in plasma NA are not solely due to a baroreflex-mediated counterregulation of the peripheral vasodilatation. Instead, they appear, at least in part, to be a consequence of ISO acting at peripheral presynaptic β-adrenoceptors. As far as the central venous plasma level of DOPEG is concerned, it tended to fall below baseline during the first infusion period. This was probably due to the observed increase in cardiac output induced by ISO (Ludwig et al., 1989). However, in each individual subject plasma DOPEG increased when the rate of ISO infusion was increased during the 2nd infusion period (Fig. 2B). The increase in the rate of ISO infusion from the 1st to the 2nd infusion period gave similar absolute increments in both, plasma DOPEG ($\triangle$DOPEG) and plasma NA ($\triangle$NA). This confirms the results obtained in study 1. Moreover, like in study 1, desipramine abolished the plasma DOPEG response (Fig. 2B), indicating that the ISO-induced increase in plasma DOPEG originates from the uptake source of DOPEG formation.

In a recent study which involved 12 normotensive controls and 12 essential hypertensive subjects, we found higher than normal plasma DOPEG levels in the hypertensive group (Ludwig et al., 1990). This finding led us to hypothesize a normotensive-hypertensive difference with respect to the presynaptic NA metabolism. Thus, we were encouraged to investigate plasma NA and DOPEG concentrations in a larger groups of normotensives and hypertensives (study 3; Ludwig et al., 1991). In accordance to study 1 all subjects were exposed to graded orthostatic stress. The measurements of plasma NA at rest and during upright posture gave similar values in both groups of subjects. However, at any given plasma NA, plasma DOPEG levels were significantly higher in hypertensive than in normotensive subjects. For both groups of subjects a plot of the mean values of plasma DOPEG (ordinate) against plasma NA (abscissa) gave linear relations

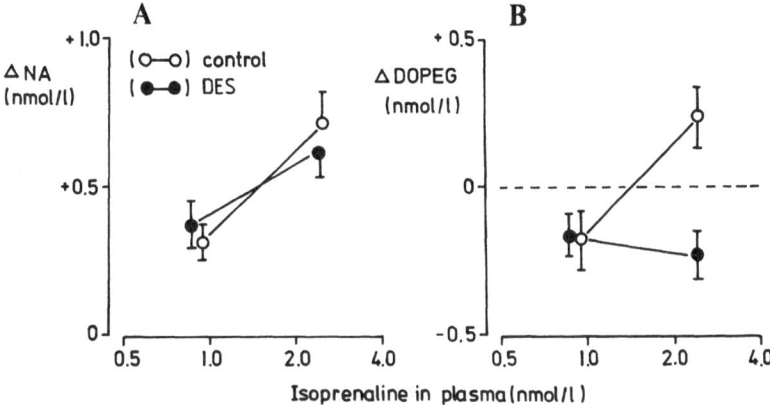

Fig. 2. Isoprenaline-induced changes in mixed central venous plasma concentration of noradrenaline (△NA; left panel) and dihydroxyphenylglycol (△DOPEG; right panel) as a function of the isoprenaline concentration (log scale) in mixed central venous plasma. Given are the results (means ± SEM) obtained under control conditions (○—○; n = 7) and after desipramine pretreatment (●—●; n = 5). Drawn from results reported by Ludwig et al. (1989)

between the two parameters. Therefore, individual regression lines were calculated in each subject to obtain individual slopes and ordinate intercepts (i.e., plasma DOPEG at zero plasma NA). It was found that the mean slope was steeper in hypertensives compared to normotensives (1.39 vs. 1.09; P < 0.02) and that the mean ordinate intercept was higher in hypertensive than normotensive subjects (4.31 vs. 2.70 nmol l^{-1}; P < 0.01). Under conditions of desipramine pretreatment it was found that the drug did not alter plasma NA responses, but abolished the plasma DOPEG responses in both groups. Moreover, the average plasma DOPEG concentration observed in the presence of desipramine was again lower in normotensives than in hypertensives (P < 0.01). As in study 1, these values did not differ from the plasma DOPEG concentrations at zero plasma NA obtained from regression analysis of the control data (see above). In conclusion, the higher than normal plasma DOPEG concentrations observed in our group of hypertensives originate from both, the uptake source and the leakage source of DOPEG formation.

In order to discuss the findings of study 3 in a meaningful way, the assumption was made that the increases in plasma DOPEG found in essential hypertension reflect an enhanced presynaptic DOPEG formation (for discussion, see Ludwig et al., 1991). One explanation could be that essential hypertension is associated with a faulty vesicular function. This would increase both the net leakage of NA from storage vesicles and the proportion of neuronally recaptured NA being transformed to DOPEG. However, this interpretation of our data requires a further assumption, namely that the vesicular dysfunction is counterbalanced by an increase in NA synthesis, since otherwise the NA stores inevitably will run empty. Another interpretation of our data is a hypertensive-normotensive difference as to the amounts of NA recaptured by the nerves and leaking out of otherwise normal storage vesicles. This would lead to the hypothesis of a hypernoradrenergic innervation of blood vessels in essential hypertension. Because, an increased nerve density is likely to correspond with an increase in available re-uptake sites as well as an increase in the number of storage vesicles. Indeed, this suggestion could help to explain both, the increase in the desipramine-resistant and desipramine-sensitive component of DOPEG formation. Moreover, the concept of hyperinnervation is supported by the finding that the spontaneous hypertension in the rat is associated with an enhanced noradrenergic innervation (Head, 1989). Interestingly, in this animal type of primary hypertension higher than normal plasma DOPEG levels were observed (Vlachakis and Alexander, 1981). Hence, although the evidence is circumstantial, our results are compatible with the concept that essential hypertension, like the spontaneous hypertension of the rat, is associated with a hypernoradrenergic innervation.

Acknowledgements

The studies of the authors reported here were supported by the Deutsche Forschungsgemeinschaft (DFG Gr 490/5) and by the Ernst und Hedda Wollheim-Stiftung, Würzburg, Federal Republic of Germany. The authors are grateful to A. Thyen for skillful technical assistance.

References

Cohen MD, Finberg J, Dibner-Dunlap M, Yuih SN, Thames MD (1990) Effects of desipramine on peripheral sympathetic nerve activity. Am J Physiol 258: R876–882

Eisenhofer G, Goldstein DS, Kopin IJ (1989) Plasma dihydroxyphenylglycol for estimation of noradrenaline neuronal re-uptake in the sympathetic nervous system in vivo. Clin Sci 76: 171–182

Esler M, Jennings G, Korner P, Willet I, Dudley F, Hasking G, Anderson W, Lambert G (1988) Assessment of human sympathetic nervous activity from measurements of norepinephrine turnover. Hypertension 11: 3–20

Goldstein DS, Horwitz D, Keiser HR, Polinsky RJ, Kopin IJ (1983) Plasma l-[^{3}H]norepinephrine, d-[^{14}C]norepinephrine, and d, 1-[^{3}H]isoproternol kinetics in essential hypertension. J Clin Invest 72: 1748–1758

Graefe K-H, Henseling M (1983) Neuronal and extraneuronal uptake and metabolism of catecholamines. Gen Pharmacol 14: 27–33

Halbrügge T, Gerhardt T, Ludwig J, Heidbreder E, Graefe K-H (1988) Assay of catecholamines and dihydroxyphenylethyleneglycol in human plasma and its application in orthostasis and mental stress. Life Sci 43: 19–26

Head RJ (1989) Hypernoradrenergic innervation: its relationship to functional and hyperplastic changes in the vasculature of the spontaneously hypertensive rat. Blood Vessels 26: 1–20

Izzo JL, Thompson DA, Horwitz D (1985) Plasma dihydroxyphenylglycol (DHPG) in the in vivo assessment of human neuronal norepinephrine metabolism. Life Sci 58: 1033–1038

Kopin IJ (1985) Catecholamine metabolism: basic aspects and clinical significance. Pharmacol Rev 37: 333–364

Ludwig J, Gerhardt T, Halbrügge T, Walter J, Graefe K-H (1988) Plasma concentrations of noradrenaline and 3,4-dihydroxyphenylethyleneglycol under conditions of enhanced sympathetic activity. Eur J Clin Pharmacol 35: 261–267

Ludwig J, Halbrügge T, Vey G, Walter J, Graefe K-H (1989) Haemodynamics as a determinant of the pharmacokinetics of and the plasma catecholamine responses to isoprenaline. Eur J Clin Pharmacol 37: 493–500

Ludwig J, Gerhardt T, Halbrügge T, Walter J, Heidbreder E, Graefe K-H (1990) Effects of nisoldipine on stress-induced changes in haemodynamics and plasma catecholamines in normotensives and hypertensives. J Human Hypertens 4: 693–701

Ludwig J, Gerlich M, Halbrügge T, Graefe K-H (1991) Plasma norepinephrine and dihydroxyphenylglycol in essential hypertension. Hypertension 17: 546–552

Szabo B, Schultheiss A (1990) Desipramine inhibits sympathetic nerve activity in the rabbit. Naunyn-Schmiedebergs Arch Pharmacol 342: 469–476

Trendelenburg U, Bönisch H, Graefe K-H, Henseling M (1980) The rate constants for the efflux of metabolites of catecholamines and phenethylamines. Pharmacol Rev 31: 179–203

Vlachakis ND, Alexander N (1981) Plasma catecholamines and their major metabolites in spontaneously hypertensive rats. Life Sci 29: 467–472

Authors' address: Dr. J. Ludwig, Medical School, University of Würzburg, Klinikstrasse 6, D-W-8700 Würzburg, Federal Republic of Germany

J Neural Transm (1991) [Suppl] 34: 77–83

Effects of imipramine and some tryptamine derivatives on the efflux of ^{3}H-5-hydroxytryptamine from rabbit platelets

R. Wölfel and K.-H. Graefe

Department of Pharmacology and Toxicology, University of Würzburg,
Federal Republic of Germany

Summary. The efflux of ^{3}H-5-hydroxytryptamine (^{3}H-5-HT) from rabbit platelets (monoamine oxidase inhibited; pretreatment with reserpine) was measured in the absence and presence of various concentrations of imipramine or a number of tryptamine derivatives. The maximum efflux-accelerating effect (E_{max}) of 5-HT and some other tryptamines (e.g., N-methyl-5-HT, 5-methoxytryptamine) far exceeded that of imipramine, whereas the E_{max} for 2-methyl-5-HT did not. It is concluded that tryptamines that are more effective in releasing ^{3}H-5-HT than imipramine have the property of being substrates of the 5-HT transporter.

Introduction

Although the properties of the 5-hydroxytryptamine (5-HT) transporter associated with the plasma membrane of blood platelets have been described in detail (Sneddon, 1973; Rudnick et al., 1983; Stahl, 1985; Given and Longenecker, 1985), the question of whether tryptamine derivatives other than 5-HT are substrates of this carrier system has so far not been given the attention it deserves. The present paper describes a method which can be used to determine whether any given compound, which has affinity to the platelet 5-HT transporter, acts as substrate or inhibitor of the transporter. The methodology employed was adopted from studies of the neuronal noradrenaline transporter (uptake$_1$). It is well established that indirectly acting sympathomimetic amines release noradrenaline from the axoplasm of adrenergic neurones through a mechanism that is Ca^{2+}-independent and readily blocked by inhibitors of uptake$_1$ (Paton, 1976; Trendelenburg, 1978; Bönisch and Trendelenburg, 1988). Indirectly acting sympathomimetic amines are all alternative substrates of uptake$_1$. By being transported themselves, they provide the carrier sites at the inner aspect of the membrane required for outward transport (release) to occur. Counterflow phenomena of this kind, which are sometimes referred to as "accelerative exchange diffusion" (Paton, 1976), have been observed also at the level of the platelet plasma membrane. Nelson and Rudnick (1979) were the first to demonstrate that the unidirectional ^{3}H-5-HT efflux out of plasma membrane vesicles obtained from porcine platelets is stimulated by unlabelled external 5-HT. Hence, not only the neuronal noradrenaline transporter, but also the platelet 5-HT transporter is capable of bringing about acceleration of efflux in response to substrates added to the external medium.

The present experiments were carried out in washed rabbit platelets in which monoamine oxidase (MAO) was blocked by preexposure of the cells to pargyline. To simplify the measurement of efflux, platelets were loaded with ^{3}H-5-HT. The experimental conditions also included pretreatment of the animals with reserpine (to deplete the endogenous 5-HT stores and to block vesicular uptake). The ^{3}H-5-HT taken up by the cells under these conditions is likely to distribute mainly into the cytoplasm rather than into the storage vesicles.

Material and methods

Rabbits were anaesthetized by i.m. injection of 0.25 ml/kg Hypnorm^{TM} (Janssen, Neuss, FRG), and blood was collected from an ear artery in plastic syringes containing trisodium citrate (final concentration 10.6 mmol/l). Platelet-rich plasma (PRP) was obtained by centrifugation of the blood (175 g, 15 min) at room temperature. The buffer used to wash and incubate the platelets had the following composition (mmol/l): NaCl 140.0, KCl 5.0, MgSO$_4$ 1.2, HEPES 10.0, glucose 10.1, prostaglandin E$_1$-α-cyclo-dextrine (PGE$_1$) 0.00003. Its pH was adjusted to 7.0 with Tris. As a rule, the buffer was supplemented with 50 μl/ml platelet-free plasma.

Preparation of washed platelets

After addition of PGE$_1$ (30 nmol/l) and pargyline (100 μmol/l), the PRP was centrifuged (1000 g, 8 min). The sedimented platelets were suspended in buffer and exposed to pargyline (100 μmol/l) for 30 min. Finally, the cells were centrifuged again (1000 g, 8 min) and resuspended in buffer to give platelet-rich buffer (PRB) which contained 0.3–0.6 10^6 platelets/μl. All steps of the preparation of PRB were done at room temperature.

Uptake experiments

After preincubation of the PRB for 5 min (37°C), the incubation with ^{3}H-5-HT was started by diluting 100 μl PRB with 900 μl buffer (37°C) containing ^{3}H-5-HT (20 nmol/l) and terminated after 15 s by rapid dilution with ice-cold buffer followed by filtration (Whatman GF/F). The radioactivity remaining on the filter was taken to reflect ^{3}H-5-HT uptake; it was corrected for the blank value obtained from PRB samples exposed throughout to 50 μmol/l imipramine (for further details, see Wölfel et al., 1988). Both preincubation and incubation were carried out in the absence and presence of various concentrations of the drug under study (6–8 concentrations per drug). Concentration-effect curves for the drug-induced inhibition of uptake were used to determine IC$_{50}$ values.

Efflux experiments

In these experiments, platelets from rabbits pretreated with reserpine (0.5 mg/kg s.c. 24 h prior to blood sampling) were used. They were loaded with ^{3}H-5-HT (30 min) by adding ^{3}H-5-HT (80 nmol/l) to the buffer used to preexpose the cells, during the preparation of PRB, to pargyline (see above). Samples of 175 μl PRB thus obtained were prewarmed (5 min, 37°C) and then diluted with 585 μl buffer either containing or not containing the drug to be tested. After 5 min of incubation (37°C), a portion of the platelet suspension was transferred into filter units (Greiner, Frickenhausen, FRG) and centrifuged for 3 s at about 5000 g. The efflux of ^{3}H-5-HT (obtained from the radioactivity in the filtrate) was expressed in % of the total radioactivity present in the whole platelet suspension (i.e., the tritium content of the cells). The efflux found in the presence of drug was always corrected for the control efflux observed in the absence of drug. Concentration-effect curves for the drug-induced acceleration of efflux were determined by using 6–7 drug concentrations that corresponded to 0.5–40 times the respective IC_{50}.

Calculations and statistics

Kinetic parameters of the concentration-effect relationships concerning both inhibition of ^{3}H-5-HT uptake and acceleration of ^{3}H-5-HT efflux were calculated by fitting a logistic function equivalent to Hill's equation to the data points of each individual experiment (Wölfel et al., 1988). The results presented here include values of IC_{50} (drug concentration producing half the maximum inhibition), EC_{50} (drug concentration producing half the maximum acceleration of efflux), E_{max} (maximum acceleration of efflux in % per 5 min) and n_H (midpoint slope of the concentration-effect curve for the drug-induced acceleration of efflux).

The results given are geometric means with SEM values or 95% confidence limits (the latter being stated in parentheses). The significance of differences between means was evaluated by analysis of variance followed by the t-test modified for multiple comparisons (Wallenstein et al., 1980).

Drugs used in the study

5-[1,2-^{3}H(N)]-Hydroxytryptamine creatinine sulphate (NEN, Dreieich, FRG); pargyline hydrochloride, imipramine hydrochloride, 5-hydroxytryptamine (5-HT) creatinine sulphate, N-methyl-5-HT oxalate, 5-methoxytryptamine hydrochloride (Sigma, München, FRG), 2-methyl-5-HT maleate (RBI, Köln, FRG); S(+)-α-methyl-5-HT (Sandoz, Basel, Switzerland); prostaglandin E$_1$-α-cyclodextrine (Schwarz Pharma, Monheim, FRG); reserpine (CIBA Pharma, Wehr, FRG).

Results

The drugs which were examined in this study are listed in Table 1. They all inhibited ^{3}H-5-HT uptake. Imipramine was by far the most potent uptake inhibitor. The IC_{50} values for the tryptamine derivatives tested here ranged from 270 to 10100 nmol/l (Table 1). When determined in separate experiments under the present experimental conditions, the K_m of the transporter

Table 1. Kinetic parameters for the drug-induced inhibition of ^{3}H-5-HT uptake by, and the drug-induced increase in ^{3}H-5-HT efflux from, rabbit platelets

Drug	^{3}H-5-HT uptake		^{3}H-5-HT efflux		
	n	IC_{50} (nmol l^{-1})	n	EC_{50}/IC_{50}	E_{max} (% 5 min^{-1})	n_H
Imipramine	6	22 (17; 29)	10	3.35 (2.65; 4.24)	7.0 (5.4; 9.2)	1.11 (0.82; 1.51)
5-HT	5	270 (212; 344)	5	2.10 (1.89; 2.34)	24.4 (16.4; 36.3)	2.01 (1.75; 2.30)
N-M-5-HT	5	520 (340; 800)	5	1.51 (1.05; 2.16)	26.1 (14.7; 46.2)	1.77 (1.23; 2.54)
S(+)-α-M-5-HT	5	620 (348; 1100)	5	1.90 (1.44; 2.50)	25.2 (15.1; 41.9)·	1.85 (1.59; 2.15)
5-MO-Tryptamine	5	9900 (7510; 13050)	5	2.01 (1.67; 2.43)	23.3 (14.5; 37.4)	1.36 (1.23; 1.50)
2-M-5-HT	5	10100 (8650; 12150)	5	2.31 (1.45; 3.66)	3.4 (1.7; 6.9)	1.17 (0.72; 1.91)

Given are geometric means (with 95% confidence limits) of n observations. IC_{50} = drug concentration producing 50% inhibition of ^{3}H-5-HT uptake; EC_{50} = drug concentration producing half the maximum increase in ^{3}H-5-HT efflux; E_{max} = maximum increase in ^{3}H-5-HT efflux expressed in % of the tritium content of the cells per 5 min: n_H = apparent Hill coefficient of the concentration-effect curve. Abbreviations: *N-M-5-HT* N-methyl-5-HT; *S(+)-α-M-5-HT* S(+)-α-methyl-5-HT; *5-MO-tryptamine* 5-methoxytryptamine; *2-M-5-HT* 2-methyl-5-HT

for ^{3}H-5-HT was very similar in value to the IC_{50} for 5-HT given in Table 1 (Wölfel, unpublished observations).

The loading with ^{3}H-5-HT (0.08 μmol/l) for 30 min resulted in a platelet ^{3}H-5-HT content of 7.4 (6.4; 8.6) pmol/10^8 cells (n = 35). Assuming a mean rabbit platelet volume of about 0.7 μl/10^8 cells (Born, 1970), the intracellular ^{3}H-5-HT concentration can be calculated to be 10.6 (9.1; 12.3) μmol/l. The efflux of ^{3}H-5-HT from the loaded cells observed in the absence of any drug amounted to 8.8 (7.0; 11.1)% per 5 min (n = 35).

Efflux rates in the presence of drugs were clearly higher than efflux rates under control conditions. The concentration-response curves shown in Fig. 1 indicate that the maximum increase in efflux induced by 5-HT, N-methyl-5-HT, S(+)-α-methyl-5-HT or 5-methoxytryptamine was much more pronounced (P < 0.01) than that produced by imipramine, while the maximum response to 2-methyl-5-HT was not. The kinetic parameters (i.e., values of EC_{50}, E_{max} and n_H) obtained by fitting Hill's equation to the results of individual efflux experiments are summarized in Table 1. In order to indicate that EC_{50} values usually exceeded IC_{50} values, the EC_{50} was expressed as the ratio of EC_{50}/IC_{50}. The concentration-effect curves for those tryptamines which were more effective in facilitating efflux than

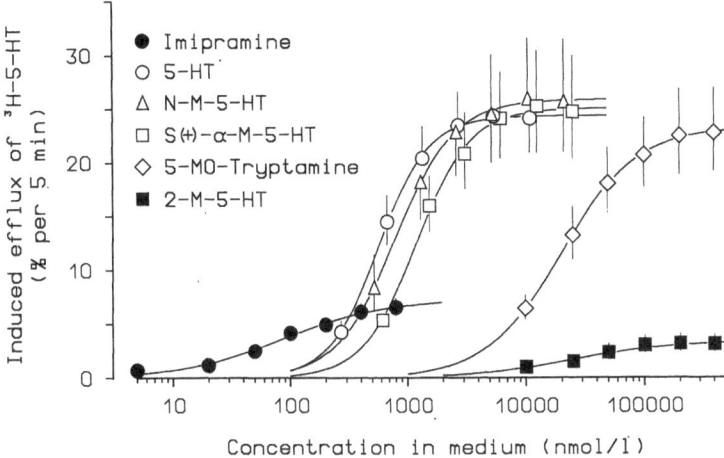

Fig. 1. Concentration-effect curves for the ³H-5-HT-releasing effects of imipramine and some tryptamine derivatives in rabbit platelets. *Ordinate*: increase in efflux of ³H-5-HT expressed in % of the tritium content of the cells per 5 min. *Abscissa*: drug concentration in the incubation medium in nmol/l. Platelets (the MAO of which was inhibited) obtained from reserpine-pretreated rabbits were first preloaded with ³H-5-HT and then exposed for 5 min to drug (for abbreviations, see Table 1) concentrations ranging from 0.5 to 40 times their IC_{50} (see Table 1). Efflux rates were corrected for the control efflux observed in the absence of any drug. Shown are geometric means ± SEM of 5 observations each (except imipramine where n = 10) and the concentration-effect curves obtained by fitting Hill's equation to the mean group data: imipramine (EC_{50} = 86.2 nmol/l; E_{max} = 7.3% per 5 min; n_H = 1.02); 5-HT (569; 24.4; 2.02); N-M-5-HT (791; 25.9; 1.72); S(+)-α-M-5-HT (1199; 25.1; 1.89); 5-MO-tryptamine (20098; 23.3; 1.36); 2-M-5-HT (24616; 3.4; 1.02)

imipramine were all characterized by Hill coefficients (n_H) greater than unity.

None of the tryptamines studied here produced significant increases in efflux when tested in the presence of 2 μmol/l imipramine (data not shown).

Discussion

The present results deal with the effects of imipramine and several tryptamine analogues (including 5-HT) on the efflux of ³H-5-HT from rabbit platelets previously loaded with the labelled amine. To ensure that most of the ³H-5-HT taken up by the cells distributes into the cytoplasm and is readily available for efflux and outward transport, rabbits were pretreated with reserpine (to block vesicular uptake), and MAO was blocked. Under these conditions 5-HT and a number of tryptamines (except 2-methyl-5-HT) were far more effective in accelerating the efflux of ³H-5-HT than imipramine.

The available evidence shows imipramine to be a pure 5-HT uptake inhibitor. It binds to the transporter protein, competitively inhibits 5-HT transport, but does not undergo transport (Talvenheimo et al., 1979). Therefore, the increase in ^{3}H-5-HT efflux produced by imipramine must solely be a consequence of re-uptake inhibition. The fact that imipramine enhanced the efflux indicates that a major component of the baseline ^{3}H-5-HT efflux is not carrier-mediated, but comes about by passive outward diffusion.

Quite in contrast to imipramine, substrates of the transporter should have the ability to induce an outward transport of ^{3}H-5-HT. This especially holds true for membrane carrier systems, such as the platelet 5-HT transporter, in which the reorientation of the empty carrier to the outer face of the membrane is rate-limiting in inward transport (Nelson and Rudnick, 1979). By being transported, substrates greatly enhance the probability of the transporter being available for outward transport. Hence, substrates accelerate the efflux not only by blocking the re-uptake of ^{3}H-5-HT, but also by inducing an outward transport of ^{3}H-5-HT. In other words, substrates should have E_{max} values in excess of that observed for imipramine. Apart from the endogenous substrate 5-HT, this was true for N-methyl-5-HT, S(+)-α-methyl-5-HT and 5-methoxytryptamine (Fig. 1). 2-Methyl-5-HT, on the other hand, did not behave as substrate and must therefore be classified as inhibitor. As far as S(+)-α-methyl-5-HT is concerned, our results are in agreement with those of Born et al. (1972) who demonstrated that this 5-HT analogue acts as a substrate of the human platelet 5-HT transporter.

In conclusion, the methodology described here may be used to characterize any given compound either as substrate or as inhibitor of the platelet 5-HT transporter. Knowledge of a possible substrate role of the various 5-HT receptor agonists is important, because the property of being a substrate would give them the latent capacity to bring about release of endogenous 5-HT and, hence, enable them to exert indirect 5-HT receptor-mediated effects.

Acknowledgements

The authors are grateful for the skilful technical assistance of A. Thyen. They are indebted to Dr. E. Schollmayer (Schwarz Pharma, Monheim, FRG) for prostaglandin E_1-α-cyclodextrine, to Dr. C. Victor (CIBA Pharma, Wehr, FRG) for reserpine and to Dr. D. Römer (Sandoz, Basel, Switzerland) for S(+)-α-methyl-5-HT.

References

Bönisch H, Trendelenburg U (1988) Mechanism of action of indirectly acting sympathomimetic amines. In: Trendelenburg U, Weiner N (eds) Catecholamines I. Springer, Berlin Heidelberg New York Tokyo, pp 246–277

Born GVR (1970) Observations on the change in shape of blood platelets brought about by adenosine diphosphate. J Physiol (London) 209: 487–511

Born GVR, Juengjaroen K, Michal F (1972) Relative activities on and uptake by human blood platelets of 5-hydroxytryptamine and several analogues. Br J Pharmacol 44: 117–139

Given MB, Longenecker GL (1985) Characteristics of serotonin uptake and release by platelets. In: Longenecker GL (ed) The platelets, physiology and pharmacology. Academic Press, Orlando San Diego New York London Toronto Montreal Sydney Tokyo, pp 463–479

Nelson PJ, Rudnick G (1979) Coupling between platelet 5-hydroxytryptamine and potassium transport. J Biol Chem 254: 10084–10089

Paton DM (1976) Characteristics of efflux of noradrenaline from adrenergic neurons. In: Paton DM (ed) The mechanism of neuronal and extraneuronal transport of catecholamines. Raven Press, New York, pp 155–174

Rudnick G, Talvenheimo J, Fishkes H, Nelson PJ (1983) Sodium ion requirements for serotonin transport and imipramine binding. Psychopharmacol Bull 19: 545–549

Sneddon JM (1973) Bood platelets as a model for monoamine-containing neurones. Prog Neurobiol 1: 151–198

Stahl SM (1985) Platelets as pharmacological models for the receptors and biochemistry of monoaminergic neurons. In: Longenecker GL (ed) The platelets, physiology and pharmacology. Academic Press, Orlando San Diego New York London Toronto Montreal Sydney Tokyo, pp 307–340

Talvenheimo J, Nelson PJ, Rudnick G (1979) Mechanism of imipramine inhibition of platelet 5-hydroxytryptamine transport. J Biol Chem 254: 4631–4635

Trendlenburg U (1978) Release induced by phenethylamines. In: Paton DM (ed) The release of catecholamines from adrenergic neurons. Pergamon Press, Oxford New York, pp 333–354

Wallenstein S, Zucker CL, Fleiss JL (1980) Some statistical methods useful in circulation research. Circ Res 47: 1–9

Wölfel R, Böhm W, Halbrügge T, Bönisch H, Graefe K-H (1988) On the 5-hydroxytryptamine transport across the plasma membrane of rabbit platelets and its inhibition by imipramine. Naunyn-Schmiedebergs Arch Pharmacol 338: 1–8

Authors' address: Dr. K.-H. Graefe, Institute of Pharmacology and Toxicology, University of Würzburg, Versbacher Strasse 9, D-W-8700 Würzburg, Federal Republic of Germany

J Neural Transm (1991) [Suppl] 34: 85–90

Influence of antidepressant drugs on seizure susceptibility and the anticonvulsant activity of valproate in mice

Z. Kleinrok, J. Gustaw, and **S. J. Czuczwar**

Department of Pharmacology, Medical School, Lublin, Poland

Summary. The tricyclic antidepressants, amitriptiline (20–30 mg/kg, i.p.) and imipramine (30–40 mg/kg), provided a significant protection against electro-convulsions (12 mA, 0.2 s stimulus duration) but desipramine (up to 40 mg/kg) remained ineffective. On the other hand, all drugs, amitriptiline (10 mg/kg), desipramine (20 mg/kg), and imipramine (20 mg/kg) distinctly potentiated the protective efficacy of valproate against maximal electroshock, reducing its ED 50 values from 255 mg/kg to 150, 135, and 128 mg/kg, respectively. In one case the plasma valproate level was measured and it was evident that desipramine (20 mg/kg) did not affect the plasma level of this antiepileptic.

Introduction

It is clearly documented that the reserpine-induced depletion of central monoamine levels leads to a considerable increase in seizure susceptibility (Chen et al., 1954). In contrast, agents enhancing monoamnine transmission (especially noradrenergic) were shown to possess anticonvulsant properties (Kilian and Frey, 1973; Löscher and Czuczwar, 1987). There are also reports indicating that some dopaminergic and serotonergic agonists may exert protective effects in certain types of experimental models of epilepsy (Janusz and Kleinrok, 1989; Löscher and Czuczwar, 1985, 1986; McKenzie and Soroko, 1972; Meldrum et al., 1975). On the basis of these data one could infer that tricyclic antidepressant drugs, which increase synaptic monoamine levels by inhibiting their re-uptake (Glowinski and Axelrod, 1964) should exhibit anticonvulsant actions. However, this was not always found to be the case. Specifically, imipramine increased the response in photosensitive baboons upon photic stimulation and in higher doses even induced spontaneous seizures (Trimble et al., 1977). Convulsive seizures were also observed during the clinical trials of imipramine, amitriptiline, desipramine, and clomipramine (for review see Trimble, 1978). Nevertheless, some experimental data report on the anticonvulsant effects of imipramine against sound- and electroshock-induced seizures (Lehmann, 1970; Maynert, 1969). Moreover, Fromm et al. (1972, 1978) provided evidence on the protective effect of imipramine on petit mal epilepsy, although a proconvulsant action on major seizures was noted.

Having in mind that seizures may be encountered in patients receiving therapeutic doses of antidepressants (Leyberg and Denmark, 1959) and because the incidence of seizures is higher when there is a risk of epilepsy or brain damage (Dallos and Heathfield, 1969), we studied the effects of some tricyclic antidepressant drugs on both the incidence of electroconvulsions and the protective efficacy of valproate against maximal electroshock.

Materials and methods

General

Experiments were performed on Swiss male mice weighing 25–30 g. The animals were housed in standard laboratory conditions with free access to food and tap water, being kept on a natural light-dark cycle.

Convulsive procedure

Electroconvulsions were produced by means of corneal electrodes and alternating current of 50 Hz, the stimulus duration being 0.2 s. The extension of the hind limbs was taken as the endpoint.

The effects of antidepressant drugs on electroconvulsions were challenged with the current intensity of 12 mA which is CS 97 (current strength necessary to produce the tonic hindlimb extension in 97% of control mice) and the protective efficacy of valproate was evaluated as its ED 50 (in mg/kg) against maximal electroshock (50 mA)-induced convulsions. The experimental groups consisted of 8–10 animals and each mouse was used only once. The convulsive test was carried out between 10.00 a.m. and 1.00 p.m.

Determination of valproate plasma level

Mice were administered either valproate alone (150 mg/kg + saline) or in combination with desipramine (20 mg/kg) and killed at appropriate times (based on the convulsive test schedule) by decapitation. Blood samples were centrifuged at 10,000 r.p.m. for 3 min and plasma samples of 70 µl were pipetted into Abboth System Assays cartridges. The plasma level of valproate was measured with the help of the Abbott TDx analyzer (Abbott, U.S.A.), operating on an immunofluorescence basis.

Drugs

The following antidepressant drugs were used in this study: amitriptiline, imipramine (both from Polfa, Poland), and desipramine (Ciba-Geigy, Switzerland). All antidepressants were in the form of hydrochloride salts. They were dissolved in sterile saline and given i.p. 45 min prior to the test.

Valproate magnesium (Polfa) was brought into solution with sterile saline and administered i.p. 30 min before maximal electroshock. Doses of antidepressant drugs and valproate refer to the free base.

Calculation of data and statistics

The protective efficacy of antidepressant drugs against electroconvulsions was expressed as the number of animals responding with tonic hindlimb extension versus the total number of mice used within an experimental group. Fisher's exact probability test was used for statistical comparisons in these cases.

The ED 50 values of valproate and their statistical evaluation were calculated according to Litchfield and Wilcoxon (1949). The original method was modified in that the computer construction of the dose-effect relationship was performed.

Plasma level of valproate was expressed in µg/ml of plasma (as a mean ± S.D. of 8 determinations) and statistical analysis of the data was carried out by means of Student's t-test.

Results

Both, imipramine (30 and 40 mg/kg) and amitriptiline (20 and 30 mg/kg) provided a considerable protection against tonic hindlimb extension. On the other hand, desipramine (up to 40 mg/kg) remained ineffective (Table 1).

All antidepressant drugs tested considerably enhanced the protective efficacy of valproate, amitriptiline (10 mg/kg) reducing valproate's ED 50 from 255 to 150 mg/kg, imipramine (20 mg/kg) — to 128 mg/kg, and desipramine (20 mg/kg) — to 135 mg/kg (Table 2).

Desipramine (20 mg/kg; 45 min before blood sampling) was without effect on the plasma level of valproate (150 mg/kg; 30 min). The control plasma level of this antiepileptic drug was 157 ± 16.3 µg/ml, whilst in combination with the antidepressant drug the level of valproate was 166 ± 20.4 µg/ml.

Table 1. Effects of tricyclic antidepressant drugs upon the incidence of tonic convulsions induced by electric current of 12 mA

Treatment (mg/kg)	No. of mice with tonic seizures	No. of mice tested
Saline	10	10
Amitriptiline (10)	9	10
Amitriptiline (20)	3*	10
Amitriptiline (30)	2**	10
Desipramine (20)	8	10
Desipramine (40)	8	10
Imipramine (20)	9	10
Imipramine (30)	4*	10
Imipramine (40)	3*	10

All drugs were administered i.p., 45 min prior to electroconvulsions. Statistical analysis of the data was performed by means of Fisher's exact probability test.
* $P < 0.01$; ** $P < 0.001$ vs. saline-treated group

Table 2. Influence of tricyclic antidepressant drugs upon the protective efficacy of valproate against maximal electroshock-induced convulsions in mice

Treatment (mg/kg)	ED 50 of valproate (mg/kg)
Saline	255 (201–318)
Amitriptiline (10)	150 (131–171)*
Desipramine (20)	135 (112–162)**
Imipramine (20)	128 (102–160)**

All drugs were given i.p.; antidepressants — 45 min and valproate — 30 min before maximal electroshock. Data are ED 50 values of valproate (in mg/kg) with 95% confidence limits in parentheses. The calculation of the ED 50 values and their statistical comparisons were made according to Litchfield and Wilcoxon (1949).
n = 8 − 10 per experimental group.
*P < 0.01; **P < 0.001 vs. control group

Discussion

The results of this study suggest that amitriptiline and imipramine, but not desipramine, provided a considerable protection against electroconvulsions in mice. It is of importance that apart from the common properties of these antidepressants, these drugs also possess N-methyl-D-aspartate (NMDA) receptor blocking effects (Reynolds and Miller, 1988). However, the protection offered against electroconvulsions was not correlated with their affinities to NMDA receptors. On the one hand, desipramine possessed the most potent MK-801 (a NMDA receptor antagonist) displacing properties, displaying no anticonvulsive activity, whilst amitriptiline was the least potent MK-801 displacer and possessed the strongest anticonvulsive activity. However, the selective NMDA receptor antagonists (phosphonic amino acids, MK-801) were not very potent against electroconvulsions (Czuczwar et al., 1984, 1985; Urbańska et al., 1991) but distinctly potentiated the protective efficacy of some antiepileptic drugs against maximal electroshock-induced seizures (Czuczwar et al., 1984; Urbańska et al., 1991). This was in fact the case when the antidepressant drugs were combined with valproate, although other mechanisms (see Introduction) than NMDA receptor blockade could participate in the potentiation of the protective activity of valproate. It is well documented that the enhancement of monoaminergic neurotransmission is associated with an increased anticonvulsive activity of a number of antiepileptic drugs (Kleinrok et al., 1980). It is also noteworthy that Fisher and Müller (1988) found a potentiation by desipramine of the protective activity of phenobarbital against maximal electroshock in mice. One should not forget that the antimuscarinic properties of antidepressants (Rehavi and Sokolovsky, 1978) might also be responsible for the enhanced effects of valproate.

Summing up, the experimental data are in disagreement with clinical observations which point to caution in the clinical use of tricyclic antidepressants, especially in patients with a positive family history of epilepsy or brain damage (Dallos and Heathfield, 1969; Leyberg and Denmark, 1959). However, the present results are based on an acute schedule and the observations from clinical studies result from chronic treatments with anti-depressants. Therefore, the effect of a prolonged treatment with anti-depressant drugs upon the seizure susceptibility and protective efficacy of antiepileptic drugs should be evaluated. Nevertheless, proconvulsant effects of imipramine were reported in primates upon a single administration, and even spontaneous convulsions were observed when the dosage of imipramine was increased (Trimble, 1978). This could also point to the possibility of different primate and rodent reactions to antidepressants. There is also a view, represented by Lange et al. (1976), that tricyclic antidepressants display anticonvulsant activity at low doses but become proconvulsant at higher dose levels. Another possibility for the pro- and convulsant activities of antidepressant drugs might be their inhibitory influence on the GABA-receptor-mediated chloride uptake (Malatynska et al., 1988). Again, one can assume that this mechanism may prevale in primates.

It is evident that tricyclic antidepressant drugs exert a variety of effects upon central neurotransmission, and initial disturbances in some neuro-transmitter systems may lead to different final influences upon seizure susceptibility.

References

Chen G, Ensor CR, Bohner BA (1954) A facilitative action of reserpine on the central nervous system. Proc Soc Exp Biol 86: 507–510

Czuczwar SJ, Turski L, Schwarz M, Turski WA, Kleinrok Z (1984) Effects of excitatory amino-acid antagonists on the anticonvulsant action of phenobarbital or diphenylhydantoin in mice. Eur J Pharmacol 100: 357–362

Czuczwar SJ, Cavalheiro EA, Turski L, Turski WA, Kleinrok Z (1985) Phosphonic analogues of excitatory amino acids raise the threshold for maximal electroconvulsions in mice. Neurosci Res 3: 86–90

Dallos V, Heathfield K (1969) Iatrogenic epilepsy due to antidepressant drugs. Br Med J 4: 80–82

Fischer W, Müller M (1988) Pharmacological modulation of central monoaminergic systems and influence on the anticonvulsant effectiveness of standard antiepileptics in maximal electroshock seizure. Biomed Biochim Acta 47: 631–645

Fromm GH, Amores CY, Thies W (1972) Imipramine in epilepsy. Arch Neurol 27: 198–204

Fromm GH, Wessel HB, Glass JD, Alvin JD, Van Horn G (1978) Imipramine in absence and myoclonic-astatic seizures. Neurology 28: 953–957

Glowinski J, Axelrod J (1964) Inhibition of uptake of tritiated noradrenaline in the intact rat brain by imipramine and structurally related compounds. Nature 204: 1318–1319

Janusz W, Kleinrok Z (1989) The role of the central serotonergic system in pilocarpine-induced seizures: receptor mechanisms. Neurosci Res 7: 144–153

Kilian M, Frey H-H (1973) Central monoamines and convulsive thresholds in mice and rats. Neuropharmacology 12: 681–692

Kleinrok Z, Czuczwar SJ, Kozicka M (1980) Effect of dopaminergic and GABA-ergic drugs given alone or in combination on the anticonvulsant action of phenobarbital and diphenylhydantoin in the electroshock test in mice. Epilepsia 21: 519–529

Lange SC, Julien RM, Fowler GW (1976) Biphasic effects of imipramine in experimental models of epilepsy. Epilepsia 17: 183–196

Lehmann AG (1970) Psychopharmacology of the response to noise with special reference to audiogenic seizure in mice. In: Welch BL, Welch AS (eds) Physiological effects of noise. Plenum Press, New York, pp 227–257

Leyberg JT, Denmark JC (1959) The treatment of depressive states with imipramine hydrochloride (Tofranil). J Ment Sci 105: 1123–1126

Litchfield JT, Wilcoxon F (1949) A simplified method of evaluating dose-effect experiments. J Pharmacol Exp Ther 96: 99–113

Löscher W, Czuczwar SJ (1985) Evaluation of the 5-hydroxytryptamine receptor agonist 8-hydroxy-2-(DI-n-propylamino)tetralin in different rodent models of epilepsy. Neurosci Lett 60: 201–206

Löscher W, Czuczwar SJ (1986) Studies on the involvement of dopamine D-1 and D-2 receptors in the anticonvulsant effect of dopamine agonists in various rodent models of epilepsy. Eur J Pharmacol 128: 55–65

Löscher W, Czuczwar SJ (1987) Comparison of drugs with different selectivity for central α_1- and α_2-adrenoceptors in animal models of epilepsy. Neurosci Res 1: 165–172

Malatynska E, Knapp RJ, Ikeda M, Yamamura HI (1988) Antidepressants and seizure-interactions at the GABA-receptor chloride-ionophore complex. Life Sci 43: 303–307

Maynert EW (1969) The role of biochemical and neurohumoral factors in the laboratory evaluation of antiepileptic drugs. Epilepsia 10: 145–162

McKenzie GM, Soroko FE (1972) The effects of apomorphine, (+)-amphetamine and L-DOPA on maximal electroshock convulsions — A comparative study in the rat and mouse. J Pharm Pharmacol 24: 696–701

Meldrum BS, Anlezark GM, Trimble M (1975) Drugs modifying dopaminergic activity and behavior, the EEG and epilepsy in Papio papio. Eur J Pharmacol 32: 203–213

Rehavi M, Sokolovsky M (1978) Multiple binding sites of tricyclic antidepressant drugs to mammalian receptors. Brain Res 149: 525–529

Reynolds IJ, Miller RJ (1988) Tricyclic antidepressants block N-methyl-D-aspartate receptors: similarities to the action of zinc. Br J Pharmacol 95: 95–102

Trimble MR (1978) Non-monoamine oxidase inhibitor antidepressants and epilepsy. Epilepsia 19: 241–250

Trimble MR, Meldrum B (1977) Seizure activity in photosensitive baboons following antidepressant drugs and the role of serotoninergic mechanisms. Psychopharmacology 51: 159–164

Urbańska E, Dziki M, Kleinrok Z, Czuczwar SJ, Turski WA (1991) Influence of MK-801 on the anticonvulsant activity of common antiepileptics. Eur J Pharmacol 200: 277–282

Authors' address: Dr. Z. Kleinrok, Department of Pharmacology, Medical School, Jaczewskiego 8, PL-20-090 Lublin, Poland

Neurotransmitter release and co-transmission

J Neural Transm (1991) [Suppl] 34: 93–98

Noradrenaline-ATP corelease and cotransmission following activation of nicotine receptors at postganglionic sympathetic axons

K. Starke, R. Bültmann, J. M. Bulloch, and **I. von Kügelgen**

Institute of Pharmacology, Freiburg i.Br., Federal Republic of Germany

Summary. In rabbit mesenteric arteries, nicotine-evoked vasoconstrictor responses were markedly reduced by prazosin, slightly reduced after desensitization by α, β-methylene ATP, and abolished by combined treatment with prazosin and α, β-methylene ATP. In guinea-pig vasa deferentia preincubated with [^{3}H]noradrenaline, nicotine elicited contractions as well as an overflow of tritium and of ATP. The contractions were greatly reduced by prazosin and abolished after additional desensitization by α, β-methylene ATP. The nicotine-induced overflow of tritium was not changed by either treatment. The overflow of ATP was decreased by prazosin but not diminished further after additional desensitization by α, β-methylene ATP. Activation of prejunctional nicotine receptors elicits a corelease of noradrenaline and ATP which leads to cotransmission in both tissues.

Introduction

Nicotine receptors were the first prejunctional receptors for that retrospectively evidence can be found in the literature: nicotine markedly increased the rate of beat of the rabbit isolated perfused heart (Dixon, 1924), and since the effect is mediated by noradrenaline release and yet the rabbit heart contains no sympathetic ganglion cells, the site of action of nicotine must have been at the terminal sympathetic axons. Activation of prejunctional nicotine receptors in the heart also releases dopamine β-hydroxylase (Jilg and Muscholl, 1980) and neuropeptide Y (Richardt et al., 1988), substances which are costored with noradrenaline in the sympathetic varicosity vesicles.

ATP is another constituent of the storage vesicle matrix (Schümann, 1958). The full significance of the costorage of noradrenaline and ATP is only now being appreciated, since it has been recognized that ATP is a cotransmitter of noradrenaline in several tissues (see Burnstock, 1990). The tissues include rabbit mesenteric arteries (von Kügelgen and Starke, 1985; Ramme et al., 1987; Bulloch and Starke, 1990) and the guinea-pig vas deferens (Fedan et al., 1981; Meldrum and Burnstock, 1983): contractile responses to electrical (neural) stimulation are partly resistant to α-adrenoceptor blockade, are reduced after desensitization of ATP receptors (P_{2X}-purinoceptors) by α, β-methylene ATP, and are blocked when α-antagonists and α, β-methylene ATP are combined.

The present study was devised in order to find out whether activation of prejunctional nicotine receptors, like stimulation by electrical pulses, also releases both noradrenaline and ATP, and whether this corelease, like electrically evoked corelease, leads to cotransmission.

Methods

Rabbit ileocolic arteries

Large branches of the ileocolic artery were mounted vertically in an organ bath where they were simultaneously perfused (2.7 ml/min) and incubated in medium. Vasoconstriction was measured as an increase in perfusion pressure. Drugs were added to the bath but not the perfusion fluid. Concentration-response curves were determined by non-cumulative agonist addition.

Guinea-pig vas deferens

Desheathed vasa deferentia were preincubated with [³H]noradrenaline and then mounted vertically and superfused (1 ml/min). Parameters measured were contraction, the outflow of total tritiated compounds and the outflow of ATP (luciferase technique). Drugs were infused into the superfusion stream.

Results and discussion

Rabbit ileocolic arteries

In initial experiments the selectivities of prazosin, used to block α_1-adrenoceptors, and α, β-methylene ATP, used to desensitize P_{2X}-receptors, were studied against the vasoconstrictor effects of exogenous noradrenaline and ATP. Prazosin 0.1 µmol/l shifted the concentration-response curve of noradrenaline to the right but did not change the curve for exogenous ATP. α, β-Methylene ATP 10 µmol/l caused strong but transient vasoconstriction. Subsequently, responses to ATP were markedly reduced whereas the concentration-response curve of noradrenaline was not changed. Hence, prazosin and α, β-methylene ATP, under the conditions used, selectively block the effects of noradrenaline and ATP, respectively.

Nicotine elicited monophasic perfusion pressure increases which faded while nicotine was still in the bath, reflecting the typical "explosive" transmitter release that is rapidly followed by desensitization of the prejunctional nicotine receptor (Löffelholz, 1970). In order to keep desensitization to a minimum, only a single concentration-response curve of nicotine was determined on each preparation. The curve was bell-shaped (Fig. 1).

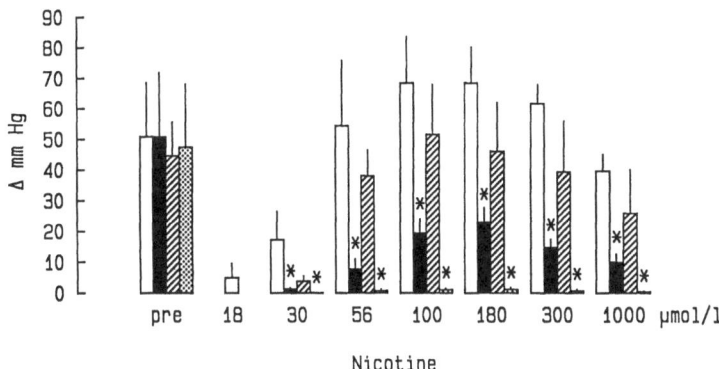

Fig. 1. Influence of prazosin and α, β-methylene ATP on the concentration-response curve of nicotine in rabbit ileocolic arteries. Nicotine 50 μmol/l was initially added three times in the absence of any other drug; the three responses were averaged (pre). Sixty minutes later, a concentration-response curve of nicotine, 30-min intervals between concentrations, was determined either in the presence of solvent (open columns), or in the presence of prazosin 0.1 μmol/l (filled columns), or after desensitization by α, β-methylene ATP 10 μmol/l (hatched columns), or after desensitization by α, β-methylene ATP 10 μmol/l with prazosin 0.1 μmol/l present as well (dotted columns). Ordinate, increase in perfusion pressure. Means ± SEM from 4–8 experiments. Significant differences from solvent (Mann-Whitney-test): *$p < 0.05$

Neither prazosin 0.1 μmol/l nor α, β-methylene ATP caused any change in the time course of the responses, but both depressed the response magnitudes. The depression was much more pronounced with prazosin than with α, β-methylene ATP. Combined treatment with prazosin and α, β-methylene ATP abolished the nicotine-induced vasoconstriction (Fig. 1). The response to nicotine was also abolished by hexamethonium 100 μmol/l.

These findings resemble observations on electrically stimulated rabbit ileocolic arteries (see Introduction). They indicate that nicotine releases both noradrenaline and ATP, and that both contribute to the vasoconstriction. The response, hence, consists of an adrenergic component (remaining in the presence of α, β-methylene ATP) and a purinergic component (remaining in the presence of prazosin). However, there are also differences to the pattern obtained with electrical stimulation. For instance, purinergic transmission predominates in rabbit mesenteric arteries and other blood vessels in the case of low frequency electrical stimulation (Kennedy et al., 1986; Ramme et al., 1987; Muramatsu et al., 1989). In the case of nicotine, in contrast, adrenergic transmission predominated throughout the range of concentrations tested.

Guinea-pig vas deferens

Can the corelease of noradrenaline and ATP also be demonstrated as a simultaneous "overflow" of the two substances? The question was studied in guinea-pig vasa deferentia preincubated with [3H]noradrenaline in

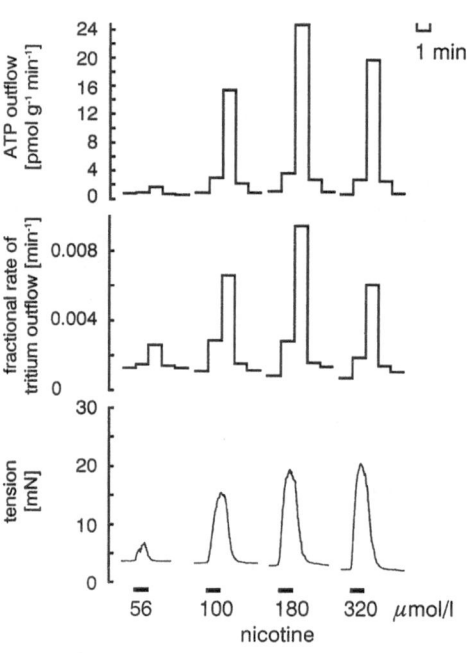

Fig. 2. Nicotine-evoked ATP overflow, noradrenaline overflow and contraction in a guinea-pig vas deferens preincubated with [^{3}H]noradrenaline. Nicotine was administered at 4 increasing concentrations for 1 min each, after 90, 150, 210 and 270 min of superfusion. The outflow of tritium is expressed as fractional rate, i.e. (outflow into superfusate in 1 min)/(tritium content of the tissue at the onset of that 1-min period). One representative experiment of four

which electrical stimulation has previously been shown to elicit an over-flow of both [^{3}H]noradrenaline and ATP (Lew and White, 1987; Kasakov et al., 1988). Nicotine produced monophasic, transient and concentration-dependent contractions which were accompanied by marked increases in the outflow of total tritiated compounds (reflecting release of [^{3}H]noradrenaline) as well as of ATP (Fig. 2). All effects were blocked by hexamethonium 100 µmol/l.

The influence of prazosin and desensitization by α, β-methylene ATP was again examined for further analysis. Nicotine 180 µmol/l was administered twice; either prazosin 0.3 µmol/l alone or prazosin 0.3 µmol/l plus α, β-methylene ATP 10 µmol/l was added 51 min before the second exposure to nicotine. As in ileocolic arteries, prazosin did not change the time course of the contractile responses but markedly (by 85% on average) reduced their magnitude; the prazosin-resistant fraction was abolished when P_{2X}-receptors were desensitized in addition. The nicotine-induced overflow of tritium was not changed by either treatment. The nicotine-induced over-flow of ATP was reduced by prazosin (by 81% on average); it was not attenuated further by combined treatment with prazosin and α, β-methylene ATP.

The prazosin-resistant component and its abolition by α, β-methylene ATP resemble the findings in rabbit ileocolic arteries. They indicate that noradrenaline and ATP are cotransmitters in nicotine-induced contractions of the guinea-pig vas deferens. Similar results were recently reported for the rat vas deferens (Carneiro and Markus, 1990). In further agreement with the rabbit arteries, responses to nicotine were monophasic, not changed in their time course by prazosin, and mainly adrenergic. Electrically evoked contractions of the guinea-pig vas deferens are often biphasic with a mainly purinergic followed by a mainly adrenergic phase (Fedan et al., 1981).

The essential finding of this series is that a nicotine-induced release of both noradrenaline and ATP can also be detected as "overflow". Since noradrenaline is stored exclusively in postganglionic sympathetic axons in the vas deferens, and since [^{3}H]noradrenaline is exclusively taken up into these axons, the source of the outflow of tritium was neuronal. ATP, however, occurs in all cells, and the source of the outflow of ATP is much less certain. It has been suggested that the electrically evoked overflow of ATP from the guinea-pig vas deferens originates entirely from sympathetic axons (Lew and White, 1987; Kasakov et al., 1988). This seems is unlikely for the overflow evoked by nicotine: in contrast to the nicotine-evoked release of [^{3}H]noradrenaline, the nicotine-evoked overflow of ATP was markedly reduced by prazosin. The fraction abolished by prazosin probably originated from smooth muscle cells, released by noradrenaline through activation of α_1-adrenoceptors. Exogenous noradrenaline also causes an α_1-adrenoceptor-mediated, presumably non-neuronal release of ATP in the guinea-pig vas deferens (Katsuragi et al., 1990). Only the prazosin-resistant fraction of the nicotine-induced overflow of ATP presumably reflected neuronal release. In agreement with a prejunctional origin, this fraction was not decreased when smooth muscle contractions were completely suppressed by additional exposure to α, β-methylene ATP.

Conclusion

Activation of prejunctional nicotine receptors, like electrical stimulation, releases both noradrenaline and ATP in rabbit ileocolic arteries and the guinea-pig vas deferens, and this corelease leads to cotransmission. There seem to be some differences, however, between the effects of electrical stimulation and of nicotine, possibly due to differential release of the two cotransmitters by the two different modes of stimulation. Nicotine elicited an overflow of ATP from the guinea-pig vas deferens of which only a minor part was neurogenic; the major part probably originated from smooth muscle cells where ATP was released by noradrenaline acting on α_1-adrenoceptors.

References

Bulloch JM, Starke K (1990) Presynaptic α_2-autoinhibition in a vascular neuroeffector junction where ATP and noradrenaline act as co-transmitters. Br J Pharmacol 99: 279–284

Burnstock G (1990) Co-transmission. Arch Int Pharmacodyn 304: 7–33

Carneiro RCG, Markus RP (1990) Presynaptic nicotinic receptors involved in release of noradrenaline and ATP from the prostatic portion of the rat vas deferens. J Pharmacol Exp Ther 255: 95–100

Dixon WE (1924) Nicotin, Coniin, Piperidin, Lupetidin, Cytisin, Lobelin, Spartein, Gelsemin. In: Heffter A (Hrsg) Handbuch der experimentellen Pharmakologie, Bd 2. Springer, Berlin, S 656–730

Fedan JS, Hogaboom GK, O'Donnell JP, Colby J, Westfall DP (1981) Contribution by purines to the neurogenic response of the vas deferens of the guinea pig. Eur J Pharmacol 69: 41–53

Jilg B, Muscholl E (1980) The use of the De Deckere-Ten Hoor preparation for study of nicotinic and potassium-evoked dopamine β-hydroxylase release from the rabbit heart. Naunyn-Schmiedebergs Arch Pharmacol 315: 139–146

Kasakov L, Ellis J, Kirkpatrick K, Milner P, Burnstock G (1988) Direct evidence for concomitant release of noradrenaline, adenosine 5'-triphosphate and neuropeptide Y from sympathetic nerve supplying the guinea-pig vas deferens. J Auton Nerv Syst 22: 75–82

Katsuragi T, Tokunaga T, Usune S, Furukawa T (1990) A possible coupling of postjunctional ATP release and transmitters' receptor stimulation in smooth muscles. Life Sci 46: 1301–1307

Kennedy C, Saville VL, Burnstock G (1986) The contributions of noradrenaline and ATP to the responses of the rabbit central ear artery to sympathetic nerve stimulation depend on the parameters of stimulation. Eur J Pharmacol 122: 291–300

Lew MJ, White TD (1987) Release of endogenous ATP during sympathetic nerve stimulation. Br J Pharmacol 92: 349–355

Löffelholz K (1970) Autoinhibition of nicotinic release of noradrenaline from postganglionic sympathetic nerves. Naunyn-Schmiedebergs Arch Pharmakol 267: 49–63

Meldrum LA, Burnstock G (1983) Evidence that ATP acts as a cotransmitter with noradrenaline in sympathetic nerves supplying the guinea-pig vas deferens. Eur J Pharmacol 92: 161–163

Muramatsu I, Ohmura T, Oshita M (1989) Comparison between sympathetic adrenergic and purinergic transmission in the dog mesenteric artery. J Physiol (Lond) 411: 227–243

Ramme D, Regenold JT, Starke K, Busse R, Illes P (1987) Identification of the neuroeffector transmitter in jejunal branches of the rabbit mesenteric artery. Naunyn-Schmiedebergs Arch Pharmacol 336: 267–273

Richardt G, Haass M, Neeb S, Hock M, Lang RE, Schömig A (1988) Nicotine-induced release of noradrenaline and neuropeptide Y in guinea pig heart. Klin Wochenschr 66 [Suppl XI]: 21–27

Schümann HJ (1958) Über den Noradrenalin- und ATP-Gehalt sympathischer Nerven. Naunyn-Schmiedebergs Arch Exp Path Pharmakol 233: 296–300

von Kügelgen I, Starke K (1985) Noradrenaline and adenosine triphosphate as co-transmitters of neurogenic vasoconstriction in rabbit mesenteric artery. J Physiol (Lond) 367: 435–455

Authors' address: Dr. K. Starke, Pharmakologisches Institut, Hermann-Herder-Strasse 5, D-W-7800 Freiburg i. Br., Federal Republic of Germany

J Neural Transm (1991) [Suppl] 34: 99–105

Preliminary evidence for noradrenaline and ATP as neurotransmitters in the porcine isolated palmar common digital artery

N. A. Blaylock and **V. G. Wilson**

Department of Physiology and Pharmacology, The Medical School, Queen's Medical
Centre, Nottingham, United Kingdom

Summary. Sympathetic neurotransmission in porcine isolated palmar common digital artery involves the release of noradrenaline and ATP which produce constrictor responses via α_1-adrenoceptor and P_{2x} receptors, respectively. Responses to short trains of pulses (e.g. 4 pulses at 2Hz) are almost entirely attributable to ATP, while those to longer trains of pulses appear to involve both transmitters. Unlike the corresponding blood vessel in man (Stevens and Mould, 1985) no evidence for "innervated" post-junctional α_2-adrenoceptors was found.

Introduction

The sympathetic control of cutaneous blood vessels has attracted much interest in recent years because of (i) recognition of their primary role in thermoregulation and (ii) the possibility that exaggerated neurogenic vascular tone contributes to peripheral circulatory conditions. In addition, these blood vessels have provided further insights into the diversity of transmitters and receptor subtypes employed by the sympathetic nervous system to exert control over blood flow.

For example, Flavahan and Vanhoutte (1986) demonstrated that neurogenic responses in the canine isolated saphenous vein were attributable to noradrenaline (NA), acting upon both α-adrenoceptor subtypes, and adenosine triphosphate (ATP), activating P_{2x} receptors. In arterial cutaneous vessels, however, not all of these receptors appear to be implicated under normal conditions. In the rabbit isolated ear artery (Kennedy et al., 1986), rabbit isolated saphenous artery (Warland and Burnstock, 1987; Dunn et al., 1991) and rat tail artery (Sneddon and Burnstock, 1984) sympathetic neurotransmission involves activation of α_1-adrenoceptors and P_{2x} receptors. Furthermore, sympathetic innervation of post-junctional α_2-adrenoceptors was only observed after abolition of α_1-adrenoceptors (rat isolated tail artery; Rajanayagam et al., 1990) or following exposure to angiotensin II (rabbit isolated saphenous artery; Dunn et al., 1991).

In the present study we have examined sympathetic neurotransmission in the porcine isolated palmar common digital artery (PCDA) because the corresponding vessel in man differs from the above arterial preparations in two respects (Stevens and Moulds, 1985). First, since α-adrenoceptor blockade abolished neurogenic responses there is no evidence to support ATP as a (co)-transmitter. Secondly, overt, neurogenic α_2-adrenoceptor-mediated responses can be demonstrated without recourse to manipulation of experimental conditions.

Methods

Porcine trotters from the thoracic limb were obtained within 1 hr of the death of the animal. The palmar common digital artery (Ghoshal and Nanda, 1975) was cleaned of adherent connective tissue and placed in pre-gassed modified Krebs-Henseleit saline containing 2% Ficoll. The vessels were stored overnight at 4°C.

Four 4 mm "ring" segments were prepared from the artery and each suspended between two wire supports (0.2 mm thick), the upper support being attached by cotton to a Grass FTO3 isometric force transducer, in a 20 ml isolated organ bath containing modified Krebs-Henseleit saline maintained at 37°C and gassed with 95% O_2, 5% CO_2. A modified glass holder encompassing platinum plate electrodes connected to an electrical stimulator was used to apply frequency dependent electrical field stimulation (EFS) to each preparation.

After 30 min equilibration, tension was slowly applied in 1 g increments until a final resting tension of 4 g was achieved. Over the next 60 min the preparations slowly relaxed to a final resting tension of 1.5–2.0 g. Repetitive stimulation of 16 pulses at 8 Hz (16/8 Hz), 0.1 ms pulse width at 25 volt, was then applied every 5–7 min until three consistent responses were obtained. Three consecutive frequency responses curves (FRCs), each separated by 60 min, were then constructed and antagonists added to the bathing solution for a minimum of 45 min between the 1st and 2nd, and 2nd and 3rd curves. Each FRC consisted of responses produced by 2 pulses at 1 Hz (2/1 Hz), 4 pulses at 2 Hz (4/2 Hz), 8 pulses at 4 Hz (8/4Hz), 16 pulses at 8 Hz (16/8 Hz), 32 pulses at 16 Hz (32/16 Hz) and 32 pulses at 1 Hz (32/1 Hz).

Unless otherwise stated, all results have been expressed as a % of the response to 32/16 Hz. Differences between means were considered statistically significant if $p < 0.05$ (Student t-test). The composition of the modified Krebs-Henseleit saline was (mM) NaCl 118.4, KCl 4.7, $CaCl_2$ 1.25, $MgSo_4 . 7H_2O$ 1.2, $NaHCO_3$ 24.9, KH_2PO_4 1.2 and glucose 11.1. The following drugs were used: phentolamine mesylate (Ciba-Geigy), YM-12617 (5-[2[[2-(ethoxyphenoxy)ethyl]amino]propyl]-2-methoxybenzene sulphonamide HCl, Yamanouchi), α,β-methylene ATP (Sigma), Ficoll 70,000 (Sigma) and tetrodotoxin (Sigma). Contractions were recorded by means of a Grass FTO3 transducer connected to a Maclab recording system.

Results

Characteristics of the motor response in the PCDA

Figure 1 shows representative trace recordings of the effect of EFS on the PCDA. Contractile responses to 2/1 Hz were observed in approximately

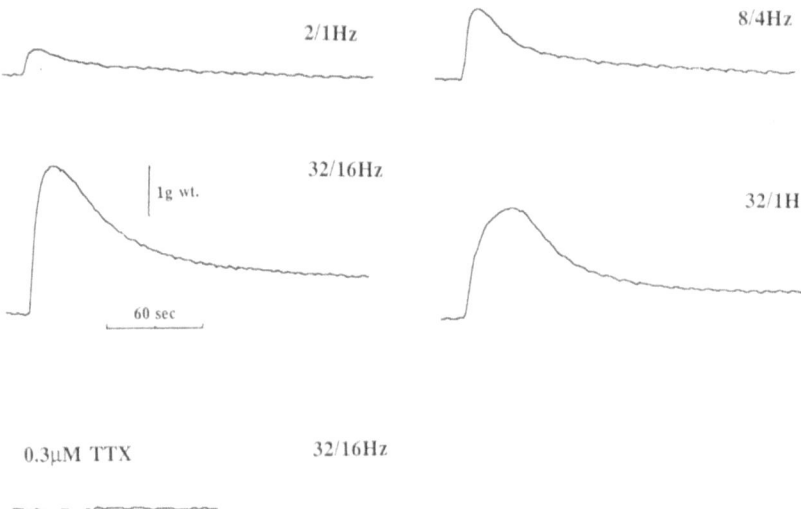

Fig. 1. Representative trace recordings of the contractile responses of the porcine isolated palmar common digital artery to electrical field stimulation

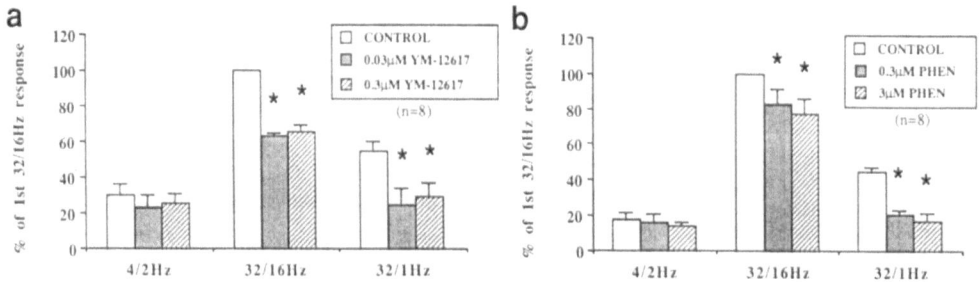

Fig. 2. The effect of **a** YM-12617 and **b** phentolamine on responses of the porcine isolated palmar common digital artery to electrical field stimulation. All responses have been expressed as a percentage of the 1st (control) response to 32/16 Hz. *Indicates a statistically significant difference ($p < 0.01$) between the responses in the presence and absence of the antagonists

1/3rd of preparations and to 4/2 Hz in virtually all preparations. These responses were characterized by a peak response within 8 s of stimulation, followed by rapid relaxation to baseline within 60 s. At higher frequencies, 8/4 Hz, 16/8 Hz and 32/16 Hz, the peak response was attained within 12 s of stimulation but was associated with an increasingly slow return to baseline, e.g. for 16/8 Hz the time to 50% relaxation, measured from the peak, was approximately 20 s and complete relaxation was usually achieved with 3 min. In contrast, stimulation at 32/1 Hz produced a response of slower onset which attained an equilibrium after 20–30 secs. On cessation of stimulation the preparations rapidly relaxed — relative to same number of pulses delivered at 16 Hz (32/16 Hz).

The maximum response to 32/16 Hz and 32/1 Hz were 89.5 ± 9.9% and 47.3 ± 5.4%, respectively of that produced by 60 mM KCl (6.05 ± 0.75 g

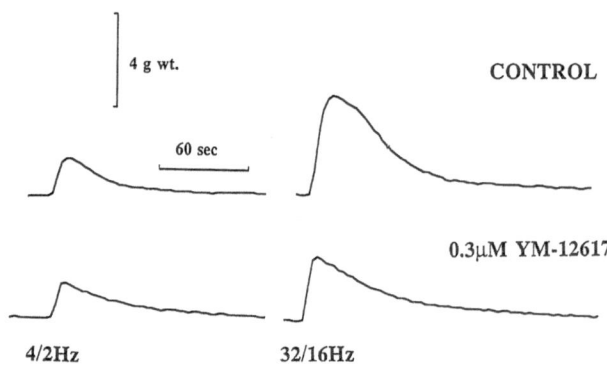

Fig. 3. Representative trace recording of contractile responses of the porcine isolated palmar common digital artery to electrical field stimulation in the absence and presence of 0.3 μM YM-12617

wt., n = 10). Frequency-responses curves (FRC) constructed at 60 min intervals were reproducibile, e.g. 32/16 Hz in the 3rd FRC was 110 ± 9.3% (n = 9) of that in the 1st FRC. All responses were abolished by the addition of 0.3 μM tetrodotoxin (n = 10).

The effect of α-adrenoceptor antagonists

Figure 2a shows the effect of the selective α_1-adrenoceptor antagonist YM-12617 (Honda et al., 1985) against electrically-evoked responses in the PCDA. 0.03 μM YM-12617 significantly reduced responses to 32/16 Hz and 32/1 Hz by approximately 40% (p < 0.01, paired t-test) but failed to affect the response to 4/2 Hz. Qualitatively similar observations were produced by 0.3 μM YM-12617. In general, YM-12617 failed to affect the rate of rise of the electrically-evoked but the duration of the response was markedly reduced (see Fig. 3). Thus, a substantial component of the response to 32/16 Hz and 32/1 Hz was unaffected by YM-12617, while the response to 4/2 Hz was completely resistant to this antagonist.

Figure 2b shows the effect of the non-selective α-adrenoceptor antagonist phentolamine (McGrath et al., 1989) on electrically-evoked responses in the PCDA. 0.3 μM phentolamine significantly reduced the responses to 32/16 Hz and 32/1 Hz by approximately 20% and 50%, respectively (p < 0.05, paired t-test), but failed to reduce the response to 4/2 Hz. A 10-fold increase in the concentration of phentolamine (3 μM) failed to produce further inhibition.

The effect of P_{2x} desensitization and α-adrenoceptor blockade

10 μM α,β-methylene ATP produced a contraction of the PCDA which relaxed spontaneously, but after 45 min, was sustained at 0.73 ± 0.24 g,

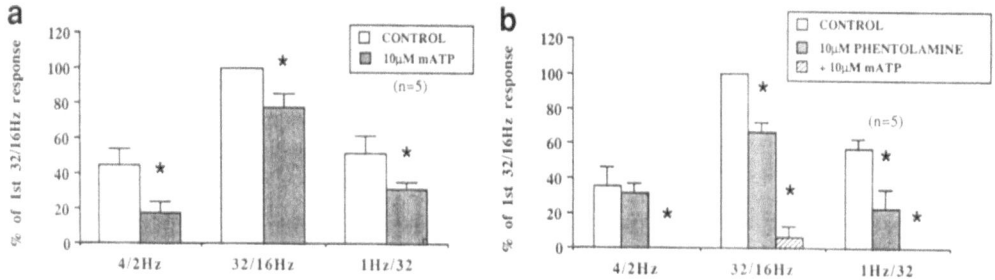

Fig. 4. The effect of **a** 10 µM α, β methylene-ATP **b** 10 µM phentolamine and 10 µM α,β methylene ATP on responses of the porcine isolated palmar common digital artery to electrical field stimulation. All responses have been expressed as a percentage of the 1st (control) response to 32/16 Hz. *Indicates a statistically significant difference (p < 0.05) between the responses in the presence and absence of the antagonists

n = 9 (equivalent to 18.9 ± 3.4% of the response to 32/16 Hz). After 45 min exposure to 10 µM α,β-methylene ATP responses to exogenous ATP were markedly reduced (not shown), response to 4/2 Hz were absent and those to 32/16 Hz and 32/1 Hz significantly reduced (Fig. 4a). The time course of responses were essentially unchanged.

Figure 4b shows the effect of 10 µM phentolamine and a combination of 10 µM phentolamine and 10 µM α,β-methylene ATP on electrically-evoked responses in the PCDA. 10 µM phentolamine reduced responses to 32/16 Hz and 32/1 Hz, but failed to reduce responses to 4/2 Hz (cf: Fig. 2b). However, the combination of 10 µM phentolamine and 10 µM α,β-methylene ATP resulted in near-complete abolition of the motor response.

Discussion

The results in the present study provide preliminary evidence that noradrenaline and ATP are co-transmitters in porcine isolated palmar common digital artery. In this respect there are clear parallels with cutaneous arteries from other species, but not man (see Introduction).

The quantitatively similar observations produced by the selective α_1-adrenoceptor antagonists YM-12617 (Honda et al., 1985) and the non-selective α-adrenoceptor antagonist phentolamine (McGrath et al., 1989), little effect against responses to a short train of pulses but 30–50% inhibition of responses to a long train of pulses (>4), suggests that the adrenergic component is solely mediated by α_1-adrenoceptors. This is further supported by the failure of low concentrations of the selective α_2-adrenoceptor antagonist rauwolscine (10 nM and 100 nM) to affect neurogenic responses, even though post-junctional α_2-adrenoceptors can be demonstrated with exogenous noradrenaline (unpublished observations). This represents a major difference from that reported in the human isolated palmar common digital artery, where both α_1- and α_2-adrenoceptors participate in electrically-evoked contractions (Stevens and Moulds, 1985).

The presence of a substantial non-adrenergic component to responses in the porcine palmar common digital artery is another difference from that in the corresponding vessel in man, being observed at all frequencies of stimulation, but particularly so for short trains of pulses (see Fig. 3). The sensitivity of this component to the P_{2x} desensitizing agent α,β-methylene ATP (Warland and Burnstock, 1987), which is only manifest after α-adrenoceptor blockade, implicates ATP as the neurotransmitter.

It is possible that these differences in sympathetic innervation between the palmar common digital artery in the pig (and cutaneous vessels in other species) and man are apparent rather than real. It is noteworthy that the stimulation protocol adopted by Stevens and Moulds (1985), continuous trains of pulses to generate a "cumulative" frequency-response curve, will almost certainly result in noradrenaline escaping the neuroeffector junction and activating "extrajunctional" α-adrenoceptors. On the other hand, short, non-cumulative trains of pulses were employed to determine the frequency-response relationship for the porcine isolated palmar common digital artery, thereby reducing the contribution of "extrajunctional" receptors to the overall response. A comparative study of the human and porcine isolated palmar common digital artery with stimulation parameters similar to those observed in vivo may resolve these discrepancies.

References

Dunn WR, McGrath JC, Wilson VG (1991) Influence of angiotensin II on the α-adrenoceptors involved in mediating the response to sympathetic nerve stimulation in the rabbit isolated distal saphenous artery. Br J Pharmacol 102: 10–12

Flavahan NA, Vanhoutte PM (1986) Sympathetic purinergic vasoconstrictor and thermosensitivity in a canine cutaneous vein. J Pharmacol Exp Ther 239: 784–789

Ghoshol NG, Nanda BS (1975) Porcine heart and arteries. In: Getty R (ed) Sissons and Grossmans' anatomy of domestic animals. Saunders and Co, Philadelphia, pp 1306–1342

Honda K, Takenaka T, Miyata-Osawa A, Terai M, Shiono K (1985) Studies on YM-12617: a selective and potent antagonist of postsynaptic α_1-adrenoceptors. Naunyn Schmiedebergs Arch Pharmacol 328: 264–272

Kennedy C, Saville VL, Burnstock G (1986) The contribution of noradrenaline and ATP to the responses of the rabbit central ear artery to sympathetic nerve stimulation depends upon on the parameters of stimulation. Eur J Pharmacol 122: 291–300

McGrath JC, Brown CM, Wilson VG (1989) α-Adrenoceptors: a critical review. Med Res Rev 9: 407–533

Rajanayagam MA, Medgett IC, Rand MJ (1990) Vasoconstrictor responses of rat tail artery to sympathatic nerve stimulation contain a component due to activation of postjunctional β- or α_2-adrenoceptors. Eur J Pharmacol 177: 35–41

Sneddon P, Burnstock G (1984) ATP as a co-transmitter in the rat tail artery. Eur J Pharmacol 106: 149–152

Stevens MJ, Moulds RFW (1985) Neuronally released norepinephrine does not preferentially activate postjunctional α_1-adrenoceptors in human blood vessels in in vitro. Circ Res 57: 399–405

Warland JJI, Burnstock G (1987) Effects of reserpine and 6-hydroxydopamine on the adrenergic and purinergic components of sympathetic nerve responses of the rabbit isolated saphenous artery. Br J Pharmacol 92: 871–880

Authors' address: Dr. V. G. Wilson, Department of Physiology and Pharmacology, Medical School, Queens Medical Centre, Clifton Boulevard, Nottingham NG 7 LUH, United Kingdom

J Neural Transm (1991) [Suppl] 34: 107–112

Involvement of catecholaminergic neurones of the nucleus of the solitary tract (NTS) in blood pressure regulation

A. Philippu, A. Pfitscher, and **N. Singewald**

Department of Pharmacodynamics and Toxicology, University of Innsbruck, Austria

Summary. Determination of the release of catecholamines in the *rostral* and *intermediate* aspects of the NTS before, during and after termination of a bilateral carotid occlusion revealed that increases in blood pressure elicited by the occlusion reduce the release rates of noradrenaline and adrenaline, while occlusion-induced decreases in blood pressure diminish the release rate of dopamine. These findings demonstrate that, in response to blood pressure changes elicited by carotid occlusion, in both aspects of the NTS noradrenaline and adrenaline act to increase blood pressure when released from their neurones, while the release of dopamine lowers blood pressure. Noradrenergic neurones of the NTS receive impulses from baroreceptors of carotid sinus and aortic arch.

Introduction

The nucleus of the solitary tract (NTS) is densely innervated by catecholaminergic neurones which seem to be involved in central cardiovascular regulation (review: Philippu, 1988). Fuxe et al. (1981) reported that noradrenergic neurones of the NTS possess a hypertensive function, while adrenergic neurones probably lower blood pressure when stimulated.

Electrical stimulation of two aspects of the NTS has been found to influence blood pressure in an opposite way; stimulation of the *rostral* aspect leads to a pressor response, while stimulation of the *intermediate* aspect of the NTS lowers blood pressure (Miura and Takayama, 1986). To investigate the importance of catecholaminergic neurones of these two aspects of the NTS in central cardiovascular control, the *rostral* NTS was superfused with artificial CSF through push-pull cannulae and the release of catecholamines was determined in the superfusate before, during and after drug-induced blood pressure changes. Moreover, the effects of a bilateral carotid occlusion on the release rates of catecholamines in the *rostral* and *intermediate* NTS were investigated.

Methods

Cats were anaesthetized with sodium pentobarbital (36 mg/kg, i.p.). Trachiotomy was carried out and both carotid arteries were prepared for bilateral occlusion. In some animals, vagus nerves, sympathetic trunks and aortic depressor nerves were transected. Catheters were inserted into the femoral artery and the femoral vein so as to record continuously mean arterial blood pressure and to inject drugs, respectively.

The head was immobilized in a stereotaxic frame. After occipital craniotomy, the cerebellum was partly sucked off thus revealing the dorsal surface of the medulla. Push-pull cannulae (Philippu et al., 1973) were inserted bilaterally into the *rostral* or *intermediate* NTS. Cannulae were stereotaxically (Berman, 1968) inserted according to the following coordinates (mm): *rostral* NTS; AP 2.7 anterior to the obex, L 2.1, V 1.4 below the surface of the medulla oblongata, *intermediate* NTS; AP 0.6 anterior to the obex, L 1.65, V 1.4 below the surface of the medulla oblongata. Cannulae had the following diameters (mm): outer cannula; outer diameter 1.27, inner diameter 1.1, inner cannula; outer diameter 0.3, inner diameter 0.1. The NTS was superfused with artificial CSF pH 7.2 (Kobilansky et al., 1988).

The superfusion rate was 150 μl/min. The superfusate was collected continuously in time periods of 3 min. Four control samples were also collected in time periods of 3 min immediately before experimentally induced blood pressure changes. Superfusates were collected in microtubes (Kobilansky et al., 1988) placed in an ice-bath and kept at −80°C until catecholamines were determined radioenzymatically (Lanzinger et al., 1989). Positions of cannulae were verified in histologic slices.

Statistical significance was calculated by Wilcoxon's signed rank-test for paired data. Mean release rates of catecholamines in the four samples preceding blood pressure changes were used as controls.

Results and discussion

Rostral aspect of the NTS

The release rates of catecholamines in the *rostral* aspect of the NTS were (fmol/min): dopamine 4.6 ± 0.9, noradrenaline 12.1 ± 1.9, adrenaline 3.9 ± 0.8 (mean values $\pm$ S. E. M., N = 13 − 20). Increases in blood pressure elicited by intravenous injections of noradrenaline, or blood re-injection, decreased the release rates of noradrenaline and adrenaline, while that of dopamine was not influenced. On the other hand, a decrease in blood pressure provoked by controlled bleeding reduced the release rate of dopamine but did not affect the release rates of noradrenaline and adrenaline. A similar effect on the release rates of catecholamines had the intravenous injection of the ganglionic blocking agent chlorisondamine which elicited a pronounced and sustained hypotension. It is obvious that increases in blood pressure lead to a counteracting inhibition of the noradrenaline and adrenaline release rates, while decreases in blood pressure inhibit the release rate of dopamine. Hence, in this aspect of the NTS noradrenaline and adrenaline possess a hypertensive function. On the other hand, dopamine seems to act hypotensive in the *rostral* aspects of the NTS (Kobilansky et al., 1988).

A bilateral carotid occlusion also changed release rates of the three

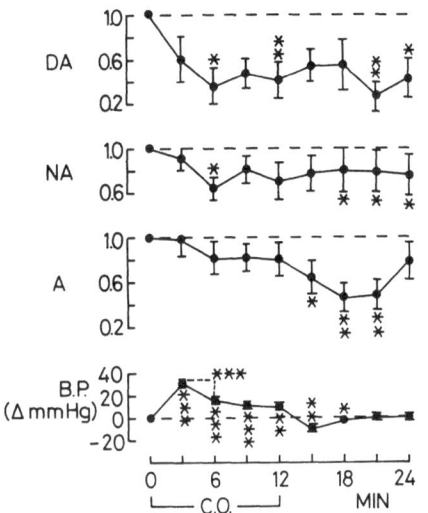

Fig. 1. Effects of blood pressure changes elicited by carotid occlusion on the release of catecholamines in the *rostral* NTS. *C.O.* carotid occlusion. Mean values ±S.E.M. of 13–20 experiments. *DA* dopamine, *NA* noradrenaline, *A* adrenaline. *P < 0.05, **P < 0.01, ***P < 0.001 (Wilcoxon). From Klausmair et al. (1991)

catecholamines in this aspect of the NTS. The release rate of dopamine was decreased during carotid occlusion and after its termination. The release rate of noradrenaline was decreased transiently during carotid occlusion. Termination of the carotid occlusion elicited a sustained inhibition in the release rate of noradrenaline. The release of adrenaline was reduced only after termination of the carotid occlusion (Fig. 1).

To explain the complex effects of the bilateral carotid occlusion on the release rates of the three catecholamines, it should be kept in mind that the NTS receives impulses from baroreceptors of the carotid sinus and of the aortic arch. During carotid occlusion, blood pressure in the carotid sinus is decreased profoundly, thus leading to a counteracting increase in the arterial blood pressure; consequently, blood pressure is also increased in the aortic arch (Table 1). Hence, during carotid occlusion firing rates of the baroreceptors of carotid sinus and aortic arch are influenced in the oposite direction. The subsequent termination of the carotid occlusion leads to a dramatic increase in blood pressure in the carotid sinus, while blood pressure in the aortic arch decreases, i.e. termination of carotid occlusion also influences differently the baroreceptors of carotid sinus and aortic arch. Based on these opposing changes in blood pressure in carotid sinus and aortic arch, alterations in the release rates of catecholamines during carotid occlusion and after its termination might be explained as follows.

1. Since increases in blood pressure inhibit the release rates of noradrenaline and adrenaline, the decreased release rates of these amines after termination of the carotid occlusion seems to be due to the increase in blood pressure in the carotid sinus.

2. The transient decrease in the release rate of noradrenaline during

Table 1. Blood pressure changes in carotid sinus and aortic arch during and after termination of carotid occlusion

	Blood pressure	
	Carotid sinus	Aortic arch
During carotid occlusion	Decrease	Increase
After termination of carotid occlusion	Increase	Decrease

During a long-lasting (12 min) carotid occlusion the pressor response declines, probably because of counteracting impulses originating from the baroreceptors of the aortic arch

carotid occlusion coincides with the statistically significant decline of the pressor response due to carotid occlusion (Fig. 1); hence, the decreased noradrenaline release might be due to counteracting impulses originating from baroreceptors of the aortic arch (see below).

3. The decreased release rate of dopamine during carotid occlusion is probably due to the low blood pressure in the carotid sinus, because the release rate of dopamine is diminished when blood pressure decreases. The decrease in the release rate of this amine after termination of the carotid occlusion might be due to impulses from baroreceptors of the aortic arch or other areas (see below).

Intermediate aspect of the NTS

In this aspect, the release rates of noradrenaline and adrenaline were comparable with those in the *rostral* aspect. The release rate of dopamine in the *intermediate* NTS was significantly higher than that in the *rostral* aspect (16.0 ± 6.2 fmol/min, N = 12; $P < 0.05$). The differing concentrations of the amine are in accordance with the differing densities of dopaminergic neurones in the two aspects of the NTS (review: Palkovits and Brownstein, 1989).

The bilateral carotid occlusion influenced the release rates of the catecholamines in a similar way, as in the *rostral* aspect of the NTS: the release rate of dopamine was reduced during and after carotid occlusion (Fig. 2), those of noradrenaline and adrenaline after termination of the occlusion. As in the *rostral* NTS, the release rate of noradrenaline was also diminished during carotid occlusion. In the *intermediate* aspect of the NTS, this change in the release rate of noradrenaline also coincided with the decline in the pressor response during carotid occlusion (Fig. 2) which is probably due to counteracting impulses originating from baroreceptors of the aortic arch. If this were so, then denervation of the aortic arch should at least reduce the inhibition in the noradrenaline release during carotid occlusion. Indeed, nerve transection abolished the decline in the release rate of noradrenaline during carotid occlusion. Surprisingly, denervation of the aortic arch did not alter the decrease in the release rate of dopamine

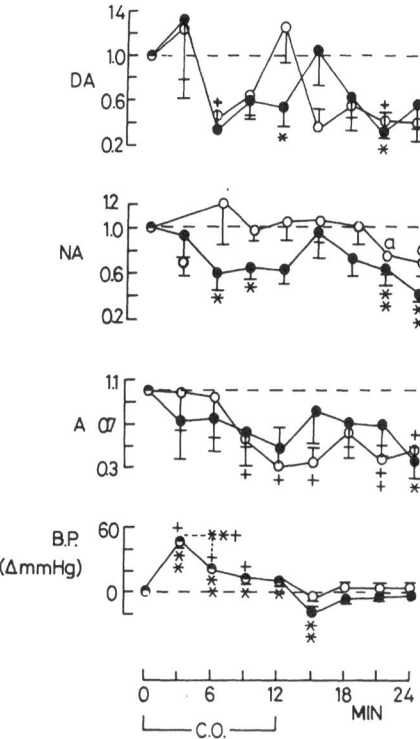

Fig. 2. Effects of blood pressure changes elicited by carotid occlusion on the release of catecholamines in the *intermediate* NTS. *C.O.* carotid occlusion. *DA* dopamine, *NA* norarenaline, *A* adrenaline. *Solid circles*: intact nerves (N = 8 − 11), *empty circles*: transected nerves (N = 6 − 14). Asterisks denote significant differences in experiments with intact nerves, crosses or α significant differences in experiments with transected nerves. * or $^+$P < 0.05, ** or $^{++}$P < 0.01, ^{α}P < 0.1 (Wilcoxon)

after termination of the carotid occlusion. It is likely that impulses from other baroreceptors are responsible for the inhibition of the dopamine release after occlusion termination (Fig. 2). It remains to be investigated, whether drug-induced changes in blood pressure influence the release of catecholamines in the *intermediate* NTS in a similar way as in the *rostral* aspect of the NTS.

These findings suggest that carotid occlusion influences in a similar way the release rates of catecholamines in the *rostral* and *intermediate* NTS. The opposing effects on blood pressure elicited by electrical stimulation of the *rostral* or *intermediate* aspects of the NTS might be due to stimulation of different catecholaminergic neurons. Furthermore, the results indicate that noradrenergic neurons of the NTS are influenced by impulses from baroreceptors of carotid sinus and aortic arch.

Acknowledgement

This work was supported by the Fonds zur Förderung der wissenschaftlichen Forschung.

References

Berman AL (1968) The brainstem of the cat. A cytoarchitectonic atlas with stereotaxic coordinates. The University of Wisconsin Press, Madison

Fuxe K, Agnati LF, Ganten D, Goldstein M, Yukimura T, Jonsson G, Bolme P, Hökfelt T, Andersson K, Härfstrand A, Unger T, Rascher W (1981) The role of noradrenaline and adrenaline neuron systems and substance P in the control of central cardiovascular functions. In: Buckley JP, Ferrario CM (eds) Central nervous system mechanisms in hypertension. Raven Press, New York, pp 89–113

Klausmair A, Singewald N, Philippu A (1991) Release of endogenous catecholamines in two different regions of the nucleus of the solitary tract as influenced by carotid occlusion. Naunyn-Schmiedebergs Arch Pharmacol 343: 155–160

Kobilansky C, Lanzinger I, Philippu A (1988) Release of endogenous catecholamines in the nucleus tractus solitarii during experimentally induced blood pressure changes. Naunyn-Schmiedebergs Arch Pharmacol 337: 125–130

Lanzinger I, Kobilansky C, Philippu A (1989) Pattern of catecholamine release in the nucleus tractus solitarii of the cat. Naunyn-Schmiedebergs Arch Pharmacol 339: 298–301

Miura M, Takayama K (1986) The functional subdivisions of the nucleus tractus solitarii of the cat in relation to the carotid sinus nerve reflex. J Auton Nerv Syst 15: 70–92

Palkovits M, Brownstein MJ (1989) Catecholamines in the central nervous system. In: Trendelenburg U, Weiner N (eds) Catecholamines II. Springer, Berlin Heidelberg New York Tokyo, pp 1–16

Philippu A (1988) Regulation of blood pressure by central neurotransmitters and neuropeptides. Rev Physiol Biochem Pharmacol 111: 1–115

Philippu A, Przuntek H, Roensberg W (1973) Superfusion of the hypothalamus with γ-aminobutyric acid: effect on release of noradrenaline and blood pressure. Naunyn-Schmiedebergs Arch Pharmacol 276: 103–118

Authors' address: Dr. A. Philippu, Department of Pharmacodynamics and Toxicology, University of Innsbruck, Peter-Mayr-Strasse 1, A-6020 Innsbruck, Austria

J Neural Transm (1991) [Suppl] 34: 113–119

Vasodilatation by endothelium-derived nitric oxide as a major determinant of noradrenaline release

T. Halbrügge[1], **K. Lütsch**[2], **A. Thyen**[1], and **K.-H. Graefe**[1]

Departments of [1]Pharmacology and [2]Pharmacy, University of Würzburg, Würzburg, Federal Republic of Germany

Summary. In the anaesthetized rabbit, L-N^G-monomethyl-arginine (L-NMMA), a specific inhibitor of nitric oxide (NO) formation, was used to assess the role of endothelium-derived NO in the regulation of haemodynamics and noradrenaline release (R_{NA}). L-NMMA dose-dependently increased mean arterial pressure and total peripheral resistance (TPR), but decreased heart rate, cardiac output and R_{NA}. The curvilinear relationship between R_{NA} and TPR obtained for L-NMMA was virtually identical with that produced by phenylephrine, indicating that L-NMMA-induced decreases in R_{NA} are mediated by the baroreflex. Since the maximum R_{NA} inhibition by L-NMMA was 69%, the counterregulation against peripheral vasodilatation by endothelium-derived NO accounts for 69% of basal R_{NA}.

Introduction

Vascular smooth muscle tone is regulated predominantly by two counteracting mechanisms: a) vasoconstriction which is dependent mainly on sympathetic tone as reflected by the release of noradrenaline and b) vasodilatation induced by endothelium-derived relaxing factor (EDRF). EDRF is considered to be identical with nitric oxide (NO) formed enzymatically from L-arginine within the endothelial cells (Palmer et al., 1988a; Schmidt et al., 1988). The L-arginine analogue L-N^G-monomethyl-arginine (L-NMMA) has been shown to act as competitive inhibitor of this NO formation from L-arginine (Palmer et al., 1988b; Sakuma et al., 1988). Furthermore, i.v. administration of L-NMMA has been reported to increase blood pressure and to decrease heart rate in the anaesthetized rabbit (Rees et al., 1989) and guinea pig (Aisaka et al., 1989). From these studies the question arises, whether a continuous basal release of NO not only leads to vasodilatation, but also modulates vascular tone by affecting noradrenaline release. Therefore, the present study aimed at examining the role of endothelium-derived NO in the regulation of haemodynamics and of noradrenaline release. L-NMMA was used to inhibit NO formation and phenylephrine to assess baroreflex-mediated changes in noradrenaline release.

Methods

Experimental procedure

Rabbits (Chinchilla bastards; 2.0–2.5 kg) of either sex were anaesthetized by i.v. injection of Saffan[R] (1:3 mixture of alfadolone and alfaxalone) and artificially respired with room air as described by Szabo et al. (1989). The left femoral artery and vein as well as the right atrium and carotid artery were cannulated with polyethylene tubing. Arterial pressure was recorded from the right carotid artery. Heart rate was obtained from the arterial pressure signal by means of a heart rate integrator. Cardiac output was determined by the thermodilution technique with a cardiac output computer.

After vessel cannulation, rabbits were infused via the left femoral vein with ^{3}H-(-)-noradrenaline ($70\,nCi\,kg^{-1}\,min^{-1}$) to determine the plasma clearance of ^{3}H-noradrenaline (see below). Blood samples for the determination of plasma levels of ^{3}H-labelled and unlabelled noradrenaline were collected from the left femoral artery (in ice-cold tubes) and centrifuged immediately.

For determination of baseline values, haemodynamic measurements were carried out and blood samples were taken 60 min after initiation of the ^{3}H-noradrenaline infusion. Thereafter, rabbits were given either L-NMMA to inhibit NO formation or infused with phenylephrine to produce α_1-adrenoceptor-mediated peripheral vasoconstriction.

In the L-NMMA group, rabbits were given three i.v. bolus injections at 10-min intervals to make up cumulative L-NMMA doses of 3, 10 and 30 (n = 7) or 10, 30 and $100\,mg\,kg^{-1}$ (n = 9). In three out of these experiments with the final L-NMMA dose of $30\,mg\,kg^{-1}$, L-arginine ($90\,mg\,kg^{-1}$) was given i.v. 10 min after the last L-NMMA injection in order to test whether the L-NMMA effects are susceptible to attenuation by L-arginine. At the end of each 10-min interval and 20 min after the L-arginine dose, haemodynamic measurements and blood sampling were repeated. L-NMMA was given in a cumulative manner, because the effects of single doses of L-NMMA have been shown to be fairly long-lasting (cf. Rees et al., 1989).

Phenylephrine was infused i.v. in two or three consecutive 10-min infusion periods at rates of 2.5 and 5 (n = 7), or 2.5, 5 and 10 (n = 9), or 5, 10 and 20 (n = 5) $\mu g\,kg^{-1}\,min^{-1}$. Haemodynamic measurements were carried out and blood samples were taken at the end of each infusion period.

Control experiments in five vehicle-treated rabbits revealed that, during a 50-min observation period, there were no time-dependent changes in any of the parameters studied.

Analytical procedure

After adsorption onto Al_2O_3 and desorption with $0.1\,mol\,l^{-1}$ $HClO_4$, plasma concentrations of noradrenaline were determined by reversed phase HPLC and electrochemical detection, with the plasma levels of endogenous noradrenaline being corrected for the analytical recovery determined in each assay (Halbrügge et al., 1988). Timed collections of the eluent leaving the HPLC system also allowed quantification of ^{3}H-noradrenaline by liquid scintillation counting. "Blank" plasma, which was spiked with known amounts ($10\,\mu l$) of the ^{3}H-noradrenaline-containing infusate, was subjected to HPLC and liquid scintillation counting as well, to determine the ^{3}H-noradrenaline infusion rate in each individual experiment (see below).

Calculations and statistics

The plasma clearance of ^{3}H-labelled noradrenaline was determined from the ratio of "^{3}H-noradrenaline infusion rate/steady-state plasma level of ^{3}H-noradrenaline" and expressed in $ml\,kg^{-1}\,min^{-1}$. Both parameters of the above ratio were obtained from the amount of radioactivity co-eluting with unlabelled noradrenaline from the HPLC system after Al_2O_3 extraction of plasma. This procedure circumvented the need to determine analytical recoveries of ^{3}H-noradrenaline. The rate of release into plasma of endogenous noradrenaline was obtained from the product of "^{3}H-noradrenaline plasma clearance x plasma level of endogenous noradrenaline" and expressed in $pmol\,kg^{-1}\,min^{-1}$. Total peripheral resistance was calculated from the ratio of "arterio-venous pressure difference/cardiac output" and expressed in $Pa\,s\,ml^{-1}$.

Dose-response curves were obtained from a least squares fit based on Hill's equation, and regression lines were calculated by the method of least squares. Results are given as arithmetic means ± SEM. To test the significance of drug-induced effects, the paired or unpaired Student's t-test was used. A P value < 0.05 was taken to indicate statistical significance.

Substances used in this study

L-N^G-Monomethyl-arginine (L-NMMA) was prepared according to the method of Corbin and Reporter (1974). The flavinate salt was converted to the hydrochloride salt by stirring with a suspension of Dowex-1 (OH$^-$ form) and titrating the resulting solution of the free base to pH 7.2 with hydrochloric acid. The identity and purity of the compound was confirmed by amino acid analysis.

Saffan^R ampoules (alfadolone:alfaxalone = 1:3, $12\,mg\,ml^{-1}$; Schweizerisches Serum and Impfinstitut, Bern, Switzerland); (-)-phenylephrine HCl, L-arginine HCl, (-)-noradrenaline bitartrate (Serva, Heidelberg, FRG); Dowex-1 (Cl$^-$ form), flavianic acid (Sigma Chemie, Deisenhofen, FRG); ^{3}H-7-(-)-noradrenaline (NET-377, $14.2\,Ci\,mmol^{-1}$) (NEN, Dreieich, FRG). The doses of drugs refer to free bases.

Results

Haemodynamics

Pretreatment baseline values of haemodynamic parameters in the L-NMMA group of animals are given in the legend to Table 1. There was no statistically significant difference between the L-NMMA and the phenylephrine group in any of the baseline values.

Both phenylephrine (2.5, 5, 10 and $20\,\mu g\,kg^{-1}\,min^{-1}$) and L-NMMA (3, 10, 30 and $100\,mg\,kg^{-1}$) dose-dependently increased mean arterial pressure and total peripheral resistance, but decreased heart rate and cardiac output. From the dose-response relationships for L-NMMA, but not from those for

Table 1. Dose-response parameters for the effects of L-NMMA on haemodynamics in anaesthetized rabbits

	ΔMAP (mm Hg)	ΔHR (min^{-1})	ΔCO (ml min^{-1})	ΔTPR (Pa s ml^{-1})
E_{max}	19.1 ± 0.6	-62.3 ± 8.8	-80.1 ± 8.8	1227 ± 373
ED_{50} (mg kg^{-1})	11.2 ± 1.1	13.4 ± 5.8	7.9 ± 2.8	18.5 ± 19.9
n_H	0.8 ± 0.1	0.8 ± 0.2	0.7 ± 0.2	0.7 ± 0.3

Results (means $\pm$ asymptotic standard error) were calculated by fitting Hill's equation to the mean group results obtained 10 min after i.v. bolus injection of cumulative L-NMMA doses of 3 (n = 7), 10 (n = 16), 30 (n = 16) and 100 (n = 9) mg kg^{-1}. Mean group baseline values ($\pm$SEM, n = 16 each) for the above parameters were as follows: mean arterial pressure (MAP) 64 ± 2 mm Hg, heart rate (HR) 285 ± 6 min^{-1}, cardiac output (CO) 336 ± 12 ml min^{-1}, and total peripheral resistance (TPR) 1457 ± 56 Pa s ml^{-1}

phenylephrine, dose-response parameters for the haemodynamic effects were obtained. The results are summarized in Table 1.

L-arginine (90 mg kg^{-1}), which has been reported to have no direct effect on either mean arterial pressure or heart rate (Rees et al., 1989), attenuated the effects of L-NMMA (30 mg kg^{-1}) on heart rate and cardiac output by about 70% and abolished the L-NMMA effects on blood pressure and total peripheral resistance.

Release of noradrenaline

In the L-NMMA group of animals (n = 16), the baseline values of plasma noradrenaline concentration and plasma clearance were 1.88 ± 0.17 nmol l^{-1} and 79.9 ± 2.0 ml kg^{-1} min^{-1}, respectively. The corresponding values obtained in the phenylephrine group (n = 21) amounted to 2.09 ± 0.16 nmol l^{-1} and 84.7 ± 2.4 ml kg^{-1} min^{-1}, respectively. Baseline values for the rates of noradrenaline release into plasma are given in the legend to Fig. 1. There was no statistically significant difference in any of these baseline values between both groups of animals.

Both phenylephrine and L-NMMA not only dose-dependently decreased the plasma noradrenaline concentration, but also decreased the noradrenaline plasma clearance. Moreover, the results obtained with both drugs fell onto a single regression line relating noradrenaline plasma clearance (ordinate) to cardiac output of plasma (abscissa). The correlation coefficient for the mean group results was r = 0.925 (n = 10).

Both phenylephrine and L-NMMA also dose-dependently reduced the rate of noradrenaline release into plasma as they increased total peripheral resistance. The curvilinear relationship between noradrenaline release and vascular resistance produced by L-NMMA was virtually identical with that produced by phenylephrine.

Dose-response curves for the % inhibition of basal noradrenaline release

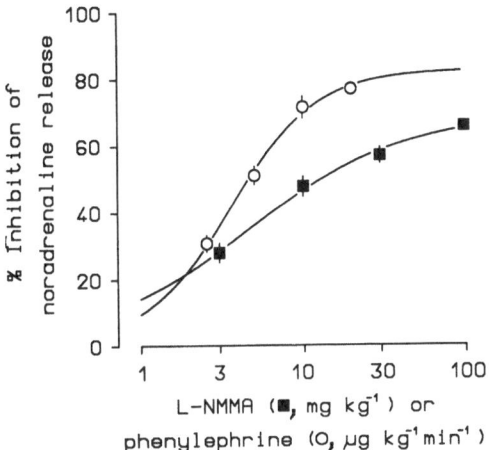

Fig. 1. Dose-response curves for the % inhibition of noradrenaline release by phenylephrine (○) or L-NMMA (■) in the anaesthetized rabbit. Pretreatment baseline values of noradrenaline release were determined prior to the administration of phenylephrine ($2.5-20 \, \mu g \, kg^{-1} min^{-1}$) or L-NMMA ($3-100 \, mg \, kg^{-1}$). Baseline noradrenaline release was $178 \pm 17 \, pmol \, kg^{-1} min^{-1}$ (n = 21) in the phenylephrine group and $152 \pm 16 \, pmol \, kg^{-1} min^{-1}$ (n = 16) in the L-NMMA group. Shown are means $\pm$SEM (for number of experiments, see text) and the dose-response curves drawn according to a fit of Hill's equation to the mean group results. The dose-response parameters obtained for phenylephrine were $E_{max} = 82.7 \pm 5.3\%$, $ED_{50} = 3.5 \pm 0.4 \, \mu g \, kg^{-1}$ and $n_H = 1.6 \pm 0.3$; the parameters for inhibition of noradrenaline release by L-NMMA were $E_{max} = 69.1 \pm 4.3\%$, $ED_{50} = 4.5 \pm 0.8 \, mg \, kg^{-1}$ and $n_H = 0.9 \pm 0.2$

by phenylephrine or L-NMMA are shown in Fig. 1. Phenylephrine showed a maximum inhibition of noradrenaline release of $82.7 \pm 5.3\%$. The blockade of NO formation by L-NMMA produced a maximum inhibition of noradrenaline release of $69 \pm 4.3\%$.

L-arginine, given in three experiments subsequent to L-NMMA (see above), attenuated the inhibitory effect of L-NMMA on noradrenaline release by 63%.

Discussion

The present results clearly confirm a physiological role for endothelium-derived NO in the control of vascular tone in vivo. Inhibition of NO synthesis by L-NMMA was found to increase blood pressure and to decrease heart rate as has been also shown by others in anaesthetized guinea pigs (Aisaka et al., 1989), rabbits (Rees et al., 1989) and rats (Gardiner et al., 1990; Rees et al., 1990). The observed L-NMMA-induced increases in blood pressure are due to a rise in total peripheral vascular resistance. This has been shown directly by measurements of regional vascular conductances in the rat (Gardiner et al., 1990) and by determination of the total peripheral vascular resistance in the rabbit (present study). However, there may be some species differences with respect to the maximum effects due to abolition of NO formation, since the attainable maximum increase in blood pressure in rats of about 46 mm Hg (Rees et al., 1990) was higher than that obtained in this study in rabbits (19 mm Hg). This difference even persists when the increases are expressed in % of baseline in order to take into account differences in pretreatment values. This is of particular significance, since the formation of NO has been reported to be dependent on the level of vascular tone (Vargas et al., 1990). Therefore, among other explanations, differences between species in the basal rate of NO synthesis or in the baroreflex sensitivity may well be considered in the interpretation of haemodynamic responses to inhibition of endothelial NO synthesis.

This study also provides evidence to show that inhibition of the L-arginine:NO pathway not only reduces endothelium-dependent vasodilatation, but also affects the neuronal vasoconstrictor component involved in the regulation of vascular tone. In our experiments the rate of noradrenaline release into plasma, but not the noradrenaline plasma concentration, was taken as a clearance-independent index of sympathetic tone, since L-NMMA as well as phenylephrine were found to cause cardiac output-dependent changes in noradrenaline plasma clearance. As shown in Fig. 1, L-NMMA dose-dependently inhibited noradrenaline release. This inhibition of release is likely to be mediated solely by the baroreflex, since the curvilinear relationship between noradrenaline release and vascular resistance produced by L-NMMA was nearly identical with that produced by phenylephrine. Moreover, the L-NMMA effects on both haemodynamics and noradrenaline release were attenuated to a similar extend by a 3-fold excess of L-arginine. From the obtained maximum inhibition of noradrenaline release by phenylephrine, it is concluded that the baroreflex-sensitive release component amounts to 83% of the basal noradrenaline release. Inhibition of NO formation by L-NMMA maximallÿreduced noradrenaline release by 69%, suggesting that the counterregulation against the EDRF-induced peripheral vasodilatation accounts for 69% of noradrenaline release under resting conditions. Hence, vasodilatation by the endogenous "nitro-vasodilator" NO plays a significant role in the regulation of noradrenaline release.

Acknowledgements

This study was supported by the Deutsche Forschungsgemeinschaft (Gr 490/5-3). The authors are grateful to Dr. O. W. Griffith for helpful advice in the preparation of L-NMMA.

References

Aisaka K, Gross SS, Griffith OW, Levi R (1989) N^G-methylarginine, an inhibitor of endothelium-derived nitric oxide synthesis, is a potent pressor agent in the guinea pig: does nitric oxide regulate blood pressure in vivo? Biochem Biophys Res Commun 160: 881–886

Corbin JL, Reporter M (1974) N^G-methylated arginines; a convenient preparation of N^G-methylarginine. Anal Biochem 57: 310–312

Gardiner SM, Compton AM, Bennett T, Palmer RMJ, Moncada S (1990) Control of regional blood flow by endothelium-derived nitric oxide. Hypertension 15: 486–492

Halbrügge T, Gerhardt T, Ludwig J, Heidbreder E, Graefe K-H (1988) Assay of catecholamines and dihydroxyphenylethyleneglycol in human plasma and its application in orthostasis and mental stress. Life Sci 43: 19–26

Palmer RMJ, Ashton DS, Moncada S (1988a) Vascular endothelial cells synthesize nitric oxide from L-arginine. Nature 333: 664–666

Palmer RMJ, Rees DD, Ashton DS, Moncada S (1988b) L-Arginine is the physiological precursor for the formation of nitric oxide in the endothelium-dependent relaxation. Biochem Biophys Res Commun 153: 1251–1256

Rees DD, Palmer RMJ, Moncada S (1989) Role of endothelium-derived nitric oxide in the regulation of blood pressure. Proc Natl Acad Sci USA 86: 3375–3378

Rees DD, Palmer RMJ, Schulz R, Hodson HF, Moncada S (1990) Characterization of three inhibitors of endothelial nitric oxide synthase in vitro and in vivo. Br J Pharmacol 101: 746–752

Sakuma I, Stuehr DJ, Gross SS, Nathan C, Levi R (1988) Identification of arginine as a precursor of endothelium-derived relaxing factor. Proc Natl Acad Sci USA 85: 8664–8667

Schmidt HHHW, Klein MM, Niroomand F, Böhme E (1988) Is arginine a physiological precursor of endothelium-derived nitric oxide? Eur J Pharmacol 148: 293–295

Szabo B, Hedler L, Starke K (1989) Peripheral presynaptic and central effects of clonidine, yohimbine and rauwolscine on the sympathetic nervous system in rabbits. Naunyn-Schmiedebergs Arch Pharmacol 340: 648–657

Vargas HM, Ignarro LJ, Chaudhuri G (1990) Physiological release of nitric oxide is dependent on the level of vascular tone. Eur J Pharmacol 190: 393–397

Authors' address: Dr. T. Halbrügge, Department of Pharmacology und Toxicology, University of Würzburg, Versbacher Strasse 9, D-W-8700 Würzburg, Federal Republic of Germany

J Neural Transm (1991) [Suppl] 34: 121–127

Stimulation of noradrenaline release in the cerebral cortex via presynaptic N-methyl-D-aspartate (NMDA) receptors and their pharmacological characterization

M. Göthert and **K. Fink**

Department of Pharmacology and Toxicology, University of Bonn,
Federal Republic of Germany

Summary. In rat brain cortex synaptosomes superfused with Mg^{2+}-free solution containing glycine, [^{3}H]noradrenaline (^{3}H-NA) release evoked by NMDA was abolished by omission of glycine or Ca^{2+} and inhibited by Mg^{2+}, the competitive and noncompetitive NMDA receptor antagonists CGP 37849 and dizocilpine (MK-801), respectively, as well as by ethanol, but was not affected by tetrodotoxin. The ^{3}H-NA release evoked by L-glutamate was also competitively inhibited by CGP 37849. In conclusion, presynaptic NMDA receptors mediate stimulation of NA release. Ethanol probably acts at this receptor system.

Introduction

Previous findings in cortical slices

Experiments in rat brain cortex slices revealed that N-methyl-D-aspartate (NMDA) and L-glutamate stimulate noradrenaline release in a Ca^{2+}-dependent manner (Fink et al., 1989). This effect was inhibited by Mg^{2+} ions, the competitive NMDA receptor antagonist 2-amino-5-phosphonovaleric acid (2-APV; Watkins et al., 1990) and dizocilpine (former designation MK-801; Lodge and Johnson, 1990); the latter drug specifically blocks the NMDA-gated cation channel by binding to the phencyclidine (PCP) recognition site inside the channel. According to this pharmacological profile we concluded that NMDA receptors mediate the stimulant effect of NMDA and L-glutamate (an endogenous agonist at these receptors). Noradrenaline release in cortical slices could also be stimulated, although at lower intrinsic activity, by kainate (Fink et al., 1989). This effect was not blocked by Mg^{2+} ions (Fink and Göthert, 1990) and, hence, was probably mediated by another class of glutamate receptors, namely the kainate receptor.

In previous investigations on peripheral noradrenergic neurones we had shown that receptors which now are classified as ligand-gated ion channels, in particular the nicotinic receptor and the 5-HT$_3$ receptor, are very sensitive to inhibition by ethanol (Göthert and Thielecke, 1976; Göthert, 1979; Göthert et al., 1979). Since the NMDA receptor and the kainate receptor also belong to the superfamily of ligand-gated ion channels, the question arose whether the noradrenaline release induced by stimulation of NMDA or kainate receptors in the *brain* is inhibited by ethanol as well. This was of interest in view of the fact that the site and mechanisms of action underlying mild to moderate intoxication by this drug are still unknown. We found ethanol at low concentrations (threshold 32 mmol/l; IC$_{50}$ 37 mmol/l) to inhibit the NMDA- or L-glutamate-evoked noradrenaline release in rat brain cortex slices, whereas the release induced by electrical stimulation, veratridine or reintroduction of Ca^{2+} after superfusion with Ca^{2+}-free K^+-rich solution was unaffected (Göthert and Fink, 1989). Ethanol also inhibited the kainate-induced noradrenaline release, although at lower potency (Fink and Göthert, 1990). Since the inhibitory effect of ethanol on NMDA-evoked noradrenaline release occurred in a concentration range which corresponds to a moderate intoxication in vivo (32 mmol/l equivalent to a blood alcohol level of 0.15%; w/v), the NMDA receptor system may be one of the sites of action via which ethanol exerts its toxic effect (Göthert and Fink, 1989).

Previous findings in synaptosomes

The exact location of the NMDA receptors involved in the effects described so far remained an open question. Since in our first study, which was carried out without exogenous glycine, NMDA failed to stimulate noradrenaline release from synaptosomes (i.e. pinched-off and resealed varicosites of nerve axon terminals), we suggested that the stimulatory NMDA receptors are not located on the noradrenergic nerve terminals but on interneurones within the slices (Fink et al., 1989). However, in the meantime the crucial role of an activation of the glycine site of the NMDA receptor system for the excitability of the NMDA receptor was established (Keith et al., 1989). Therefore, our data obtained in superfused thin layers of synaptosomes had to be reinterpreted in terms of the lack of endogenous glycine in the biophase of synaptosomal receptors: in synaptosomes any endogenous compound is washed away so rapidly that its concentration at corresponding synaptosomal recognition sites is subthreshold.

Taking these considerations into account, we found recently that in the presence of exogenous glycine, both NMDA and L-glutamate stimulated the noradrenaline release from cortical synaptosomes (Fink et al., 1990). The characterization of the receptors involved with respect to their pharmacological properties and ionic requirements for their function was performed with the natural agonist L-glutamate. Since the L-glutamate-evoked release was Ca^{2+}-dependent and was inhibited by Mg^{2+} or dizocilpine, it was concluded that NMDA receptors were involved in this effect (Fink et al., 1990), although this conclusion needed confirmation by experiments with a competitive NMDA receptor antagonist. On the other hand, a Mg^{2+}- and dizocilpine-resistant component of L-glutamate-evoked noradrenaline release was also detectable indicating that kainate receptors are also involved in noradrenaline release (Fink et al., 1990).

Aims

The main purpose of the present investigation on synaptosomes was to provide further evidence that the presynaptic receptors involved in NMDA-evoked noradrenaline release belong to the NMDA receptor class. For this purpose, NMDA itself was applied as an agonist, and the effects of modified ionic conditions as well as relevant receptor antagonists were studied. Furthermore, in supplement to our previous data with L-glutamate, we examined whether a competitive NMDA receptor antagonist produced a rightward shift of the concentration-response curve for L-glutamate-evoked noradrenaline release. Finally, we studied whether the ability of ethanol to inhibit the NMDA-evoked noradrenaline release can also be observed in synaptosomes.

Material and methods

Drugs used

L-[2, 5, 6-^{3}H]-noradrenaline ([^{3}H]NA; NEN, Dreieich, FRG); N-methyl-D-aspartic acid (NMDA), tetrodotoxin (TTX), glycine sodium salt, L-glutamate acid monosodium salt (Sigma Chemical Co., St. Louis, MO, USA); dizocilpine hydrogen maleate (MK-801; Research Biochemicals Inc., Natick, MA, USA); DL-(E)-2-amino-4-methyl-5-phosphono-3-pentanoic acid (CGP 37849; Ciba-Geigy Ltd., Basel, Switzerland); ethanol (Merck, Darmstadt, FRG).

Methods

Synaptosomes were prepared according to Mulder et al. (1975) from the brain cortex of male Wistar rats (200–300 g body weight). Details of the methods, including slight modifications of the preparation of synaptosomes, preincubation with [^{3}H]noradrenaline (43.7 Ci/mmol), superfusion of the synaptosomes and composition of the superfusion fluid have been described previously (Fink et al., 1989). At the end of the preparation procedure, 100 µl aliquots of the suspension of the synaptosomes were distributed on Whatman GF/B filters in the superfusion chambers. Subsequently, the synaptosomes were superfused for 60 min at a flow rate of 0.6 ml/min, and the superfusate was collected in 4-min samples. At the end of the experiments, the radioactivity of the superfusate samples and of the synaptosomes (radioactivity extracted with HCl 0.1 mol/l) was determined by liquid scintillation counting. Tritium overflow evoked by NMDA was calculated by subtraction of basal from total efflux during stimulation and the following 6 min and was expressed as percent of synaptosomal tritium at the onset of stimulation. Basal efflux was assumed to decline linearly from the collection period before stimulation to that 8–12 min after onset of stimulation.

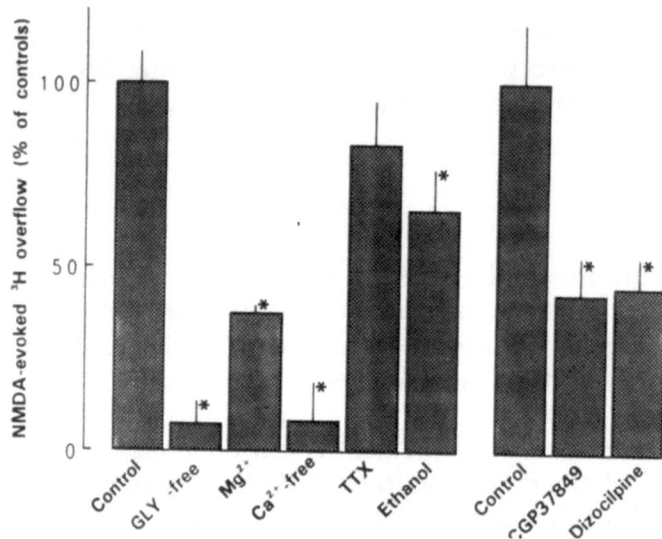

Fig. 1. NMDA-evoked tritium overflow (above basal efflux) from rat brain cortex synaptosomes preincubated with [³H]noradrenaline and superfused under various experimental conditions (including interaction with various drugs). Unless stated otherwise, the superfusion medium was Mg^{2+}-free and contained 10 µmol/l glycine (GLY) from the 20th min of superfusion onward until the end of the experiments. NMDA was added to the medium for 2 min after 40 min of superfusion. *Left group* of 6 columns: ³H overflow evoked by 1 mmol/l NMDA in the *control* experiments 0.49 ± 0.04% of synaptosomal ³H (corresponding to 0.05 ± 0.004 nCi; n = 8); *GLY-free*, experiments without glycine (n = 5); Mg^{2+}, experiments in the presence of 1.2 mmol/l $MgSO_4$ throughout superfusion (n = 7); Ca^{2+}-free, superfusion without Ca^{2+} throughout (n = 8); *TTX* (tetrodotoxin 0.3 µmol/l; n = 6) and *Ethanol* (32 mmol/l; n = 11) from the 20th min of superfusion onward until the end of the experiments. *Right group* of 3 columns: ³H overflow evoked by 100 µmol/l NMDA; ³H overflow in the *control* experiments 0.49 ± 0.08% of synaptosomal ³H (corresponding to 0.05 ± 0.01 nCi; n = 6); *CGP 37849* (30 µmol/l; for chemical name, see "Drugs used"; n = 6) and *Dizocilpine* (MK-801 0.1 µmol/l; n = 6) from the 20th min of superfusion onward until the end of the experiments. Means ± SEM. *p < 0.01 (compared with the corresponding control value)

Means ± SEM of n experiments (each in quadruplicate) are given. For comparison of two or more means, Student's t-test for paired data or Dunnett's test were used, respectively. The pA_2 value of CGP 37849 was calculated according to the formula for competitive antagonism given by Furchgott (1972; equation 4).

Results

In cortical synaptosomes superfused with Mg^{2+}-free solution containing 10 µmol/l glycine, the basal efflux of tritium under control conditions (in the fraction of superfusate collected immediately before stimulation with NMDA) was 1.28 ± 0.03% of synaptosomal tritium/min, corresponding to 0.3 ± 0.02 nCi/min; n = 6). This value was not changed by the drugs or

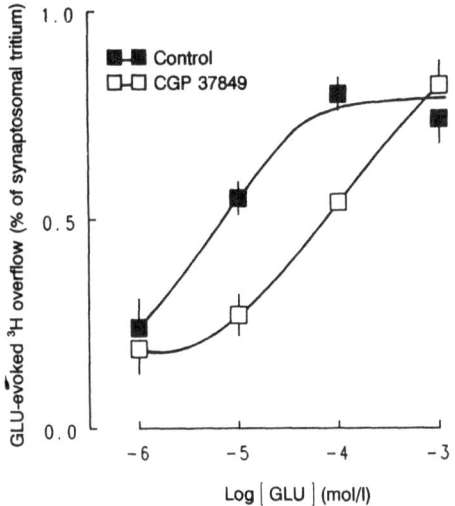

Fig. 2. L-Glutamate (L-GLU)-evoked tritium overflow (above basal efflux) from superfused rat brain cortex synaptosomes preincubated with [³H]noradrenaline, and interaction with CGP 37849. The superfusion medium was Mg^{2+}-free and contained 10 μmol/l glycine from the 20th min of superfusion onward. L-Glutamate was added to the medium for 2 min after 40 min of superfusion. Control experiments without interacting drug: ■—■; absolute value of ³H overflow evoked by, e.g., 100 μmol/l L-GLU: 0.13 ± 0.01 nCi. Experiments with CGP 37849 30 μmol/l (□—□), present in the medium from the 20th min of superfusion onward until the end of the experiments. Means ± SEM of 6 experiments. *p < 0.05 (compared with the corresponding control value obtained with the same NMDA concentration)

modifications of the composition of the superfusion medium applied in this study (results not shown; for details concerning the experimental conditions, see legends to Figs. 1 and 2).

The tritium overflow (above basal efflux) evoked by NMDA at a concentration of 1 mmol/l, which was even 10 times higher than that which produced the maximum effect, was virtually abolished by omission of glycine or Ca^{2+} and was strongly inhibited by 1.2 mmol/l Mg^{2+} (Fig. 1); tetrodotoxin was ineffective, whereas 32 mmol/l ethanol inhibited the evoked tritium overflow by 35%. In synaptosomes in which tritium overflow was evoked by 100 μmol/l NMDA, 30 μmol/l CGP 37849 or 0.1 μmol/l dizocilpine produced an inhibition by more than 50% (Fig. 1).

L-Glutamate evoked a concentration-dependent tritium overflow from cortical synaptosomes (EC_{50}: 3.3 μmol/l; Fig. 2). CGP 37849 shifted the concentration-response curve for glutamate to the right (pA_2: 5.46).

Discussion

The present data obtained in cortical synaptosomes preincubated with [³H]noradrenaline provide evidence that the NMDA-evoked tritium overflow (reflecting release of tritiated and unlabelled noradrenaline) is mediated by receptors which exhibit the characteristic features of the NMDA receptor system (for references concerning these characteristics, see Introduction): (1) The release is Ca^{2+}-dependent and sensitive to blockade by Mg^{2+} ions. (2) Activation of the modulatory glycine site is a prerequisite for the stimulant effect of NMDA. (3) The response to NMDA is inhibited by CGP 37849, a competitive NMDA receptor antagonist (Fagg et al., 1989), and by dizocilpine (MK-801), a specific blocker of the NMDA-gated cation channel (Lodge and Johnson, 1990). The failure of tetrodotoxin to block the NMDA-evoked noradrenaline release is consistent with the interpretation that this release is triggerred by cation influx via the specific ligand-gated ion channel only, whereas the fast Na^+ channel does not play a significant role.

The suggestion based on our previous data that L-glutamate also stimulates noradrenaline release via presynaptic NMDA receptors (Fink et al., 1990) is supported by the ability of CGP 37849 to produce a parallel rightward shift of the concentration-response curve for glutamate (Fig. 2). However, the pA_2 value of CGP 37849 against L-glutamate was lower than that against NMDA (unpublished result). One possible explanation for this finding is that L-glutamate, in addition to activating the NMDA receptor, may simultaneously stimulate another glutamate receptor class resistant to CGP 37849, namely the kainate receptor; this conclusion had already been drawn from our previous results as well (Fink and Göthert, 1990; Fink et al., 1990).

Since ethanol at the rather low concentration of 32 mmol/l (corresponding to a blood alcohol level of 0.15%, w/v) inhibited the NMDA-evoked noradrenaline release from cortical synaptosomes, this effect, which was first observed in cortical slices (Göthert and Fink, 1989), is obviously due to a presynaptic site of action. On the other hand, ethanol at up to 100 mmol/l did not alter the K^+ (15 mmol/l)-induced noradrenaline release from synaptosomes (unpublished results). These findings are in agreement with our previous suggestion (Göthert and Fink, 1989) that the presynaptic NMDA receptor system itself is the site of action underlying the inhibition of noradrenaline release.

Taken together, the superfused synaptosomes represent a preparation suitable to study NMDA receptor-mediated responses in a more refined system than the slice preparation. The synaptosomes still are endowed with regulatory systems relevant in situ, such as other presynaptic receptors and signal transduction mechanisms which may mutually interact with the NMDA receptor system. These mechanisms thus are accessible to systematic investigations.

Acknowledgements

This study was supported by the Deutsche Forschungsgemeinschaft and by the Dr. Robert Pfleger foundation.

References

Fagg GE, Pozza MF, Olpe HR, Brugger F, Baumann P, Bittiger H, Schmutz M, Angst C, Brundish D, Allgeier H, Heckendorn R, Dingwall JG (1989) CGP 37849: characterization of a novel and potent competitive N-methyl-D-aspartate (NMDA) receptor antagonist. Br J Pharmacol 97: 582P

Fink K, Göthert M (1990) Inhibition of N-methyl-D-aspartate-induced noradrenaline release by alcohols is related to their hydrophobicity. Eur J Pharmacol 191: 225–229

Fink K, Göthert M, Molderings G, Schlicker E (1989) N-Methyl-D-aspartate (NMDA) receptor-mediated stimulation of noradrenaline release, but not of other neurotransmitters, in the rat brain cortex: receptor location, characterization and desensitization. Naunyn-Schmiedebergs Arch Pharmacol 339: 514–521

Fink K, Bönisch H, Göthert M (1990) Presynaptic NMDA receptors stimulate noradrenaline release in the cerebral cortex. Eur J Pharmacol 185: 115–117

Furchgott RF (1972) The classification of adrenoceptors (adrenergic receptors). An evaluation from the standpoint of receptor theory. In: Blaschko H, Muscholl E (eds) Handbook of experimental pharmacology, vol 33. Springer, Berlin Heidelberg New York, pp 283–335

Göthert M (1979) Modification of catecholamine release by anaesthetics and alcohols. In: Paton DM (ed) The release of catecholamines from adrenergic neurons. Pergamon Press, Oxford New York, pp 241–261

Göthert M, Thielecke G (1976) Inhibition by ethanol of noradrenaline output from peripheral sympathetic nerves: possible interaction of ethanol with neuronal receptors. Eur J Pharmacol 37: 321–328

Göthert M, Fink K (1989) Inhibition of N-methyl-D-aspartate (NMDA)- and L-glutamate-induced noradrenaline and acetylcholine release in the rat brain by ethanol. Naunyn–Schmiedebergs Arch Pharmacol 340: 516–521

Göthert M, Dührsen U, Rieckesmann JM (1979) Ethanol, anaesthetics and other lipophilic drugs preferentially inhibit 5-hydroxytryptamine- and acetylcholine-induced noradrenaline release from sympathetic nerves. Arch Int Pharmacodyn Ther 242: 196–209

Keith RA, Mangano TJ, Meiners BA, Stumpo RJ, Klika AB, Patel J, Salama AI (1989) HA-966 acts at a modulatory glycine site to inhibit N-methyl-D-aspartate-evoked neurotransmitter release. Eur J Pharmacol 166: 393–400

Lodge D, Johnson KM (1990) Noncompetitive excitatory amino acid receptor antagonists. Trends Pharmacol Sci 11: 81–86

Mulder AH, van den Berg WB, Stoof JC (1975) Calcium-dependent release of radiolabeled catecholamines and serotonin from rat brain synaptosomes in a superfusion system. Brain Res 99: 419–424

Watkins JC, Krogsgaard–Larsen P, Honoré T (1990) Structure-activity relationship in the development of excitatory amino acid receptor agonists and competitive antagonists. Trends Pharmacol Sci 11: 25–33

Authors' address: Dr. M. Göthert, Department of Pharmacology and Toxicology, University of Bonn, Reuterstrasse 2b, D-W-5300 Bonn 1, Federal Republic of Germany

J Neural Transm (1991) [Suppl] 34: 129–137

Presynaptic modulation of neurotransmitter release by endogenous angiotensin II in brown adipose tissue

L. A. Cassis and **L. P. Dwoskin**

Division of Pharmacology and Experimental Therapeutics, College of Pharmacy,
University of Kentucky, Lexington, KY, U.S.A.

Summary. Angiotensin II (AII) increased the evoked release of [3H]-norepinephrine (NE) from superfused slices of interscapular fat (ISF). To determine if AII was endogenously formed and subsequently released from ISF, immunoreactive AII was measured in the superfusate from ISF slices. The concentration of AII detected in the ISF superfusate was 4.51 pg/mg tissue wet wt/30 ml collected over a 30-min period. In response to electrical field stimulation, AII concentration in the superfusate increased (maximum of 2-fold). To determine if AII modulates sympathetic neurotransmission, the effect of AII (0.1–10 nM) and, in separate experiments the effect of the AII-1 receptor antagonist DuP 753 (1 nM–1 μM) on the evoked release of [3H]-NE were examined in ISF slices. AII and DuP 753 increased (100% above control) and decreased (43% of control), respectively, the evoked [3H]-NE release from ISF slices. The effect of DuP 753 was not altered by the inclusion of neuronal uptake inhibitors (nomifensine or desipramine) in the superfusion buffer. These results suggest that endogenous AII enhances the evoked release of [3H]-NE from ISF.

Introduction

The major function of brown adipose tissue is the generation of heat in response to sympathetic nerve stimulation (Foster and Frydman, 1978; Cannon and Nedergaard, 1985). Brown adipose tissue is very richly innervated by the sympathetic nervous system (Cottle and Cottle, 1969; Derry et al., 1969; Foster et al., 1982; Cottle et al., 1985) with innervation to both individual brown adipocytes and blood vessels. In addition, brown adipose tissue is innervated by nerves containing a plurality of neuropeptides (Lever et al., 1986; Norman et al., 1988). We have recently demonstrated the presence of high levels of angiotensinogen messenger RNA (Ao mRNA) in brown adipose tissue of the rat (Cassis et al., 1988a,b). Levels of Ao mRNA in brown adipose tissue were as much as 60% of those in liver, the primary source of circulating angiotensinogen. Angiotensinogen serves as the only known precursor to angiotensin II (AII) (Menard et al., 1983), a peptide involved in blood pressure homeostasis; however, there is no evidence suggesting that angiotensinogen is processed to the angiotensin peptides in brown adipose tissue. Furthermore, the physiological role of AII is not known in brown adipose tissue. Presynaptic AII receptors have been demonstrated on a variety of sympathetic vascular beds (for reviews see Starke, 1977; Westfall, 1977). At subnanomolar concentrations, AII will act at these receptors to facilitate the release of NE in response to a depolarizing stimulus. In this study, the ability of endogenous and exogenous AII to modulate sympathetic neurotransmission was examined in the interscapular fat pad (ISF) of the rat, a well defined source of brown adipose tissue.

Material and methods

Animals

Male Sprague Dawley rats (250–300 g) were obtained from Harlan Laboratories (Indianapolis, Indiana). Rats were housed in groups of 2 per cage in a controlled environment under a 12 h light/dark cycle. Rats were given free access to water and food.

Tissue preparation

Rats were sacrificed by decapitation and the ISF was dissected on ice. The entire ISF (both right and left lobes) was removed, cleaned of adhering white adipose tissue and skeletal muscle, and the medial portion of each lobe was used to obtain slices. Slices of 500 μm thickness (20 mg) were obtained with a McIlwain tissue chopper. ISF slices were incubated for 30 min in a modified Krebs buffer containing: (in mM) 130 NaCl, 4.7 KCl, 1.18 KH$_2$PO$_4$, 1.17 MgSO$_4$, 14.9 NaHCO$_3$, 5.5 dextrose, 0.026 disodium EDTA, 0.11 L-ascorbic acid, 1.6 CaCl$_2$ (pH 7.4), and metanephrine (0.4 μM, to inhibit extraneuronal uptake of NE), and saturated with 95% O$_2$/5% CO$_2$ in a metabolic shaker at 37°C.

[3H]-NE release

After preincubation, each slice was incubated for an additional 30 min in the presence of 0.01 μM 1-[7-3H]-NE (specific activity 13.8 Ci/mmol, New England Nuclear, DE). Slices were then transferred to plexiglas superfusion chambers (1 ml volume) containing 2 platinum electrodes. Slices were superfused at a rate of 1 ml/min with the Krebs buffer. When included, nomifensine (NOMI, 10 μM) or desipramine (DESI, 0.5 μM) were present in the superfusion buffer at the start of superfusion. Sample (5 min intervals) collection began after 90 min of superfusion, when basal outflow of tritium stabilized to a low constant level. The first period of electrical stimulation (S1) was applied 110 min after the beginning of superfusion, and the second period of electrical stimulation (S2) was applied 85 min after S1. DuP 753 (1 nM–1 μM) or AII (0.1–10 nM) were included in the superfusion buffer 30 min before S2 and throughout the remainder of the experiment. Electrical stimulation (unipolar pulses) was delivered with a Grass stimulator (model S6C, Quincy, MA) at 30 volts, 3 msec duration, 5 Hz for 5 min. To determine the [3H] content of the tissue at the end of the experiment, each slice was solubilized, and the amount of radioactivity in each superfusate and tissue sample was determined by liquid scintillation counting. The efficiency of counting was 67%. Basal outflow and stimulation-evoked overflow were calculated as previously described (Dwoskin and Zahniser, 1986). The effect of DuP 753 or AII was assessed by comparing the evoked [3H]-overflow in response to S1 (absence of drug) to the evoked [3H]-overflow in response to S2 (presence of drug). A ratio of S2/S1 was calculated for each slice. Each experiment included a slice which was not exposed to DuP 753 or AII and served as a within experiment control.

[3H]-NE uptake

Slices of ISF were prepared as described and preincubated for 60 min in the Krebs buffer containing metanephrine (4 µM) and pargyline (1.0 µM, to inhibit monoamine oxidase) in a metabolic shaker. The time course for the [3H]-NE uptake was determined by addition of [3H]-NE (1 nM) for 5, 10, 20, 60, or 90 min. Subsequently, slices were placed in an excess of ice-cold buffer, blotted, weighed, and solubilized. A linear increase in [3H]-NE uptake was demonstrated for up to 60 min, when uptake started to plateau. Subsequently, the effect of DESI (0.1–10 µM) and NOMI (1–100 µM) on [3H]-NE uptake was determined after 10 min of incubation with [3H]-NE. After incubation, slices were placed in excess cold buffer, blotted, weighed, and solubilized. After adding cocktail, both solubilized slices and media samples were counted in a liquid scintilation counter.

Immunoreactive AII release

Slices of ISF as described above were prepared, preincubated for 30 min, and superfused for 6 h. Two periods of electrical stimulation (10 Hz, 5 min) were applied at 60 min and 270 min. Superfusate from the slices was collected over a Sep Pak C18 column (Waters, MA) which had been preequilibrated with 4 mls methanol, 4 mls of water, and 10 mls of Krebs buffer. Columns were exchanged every 30 min over the 6 h superfusion period. After washing the columns with 10 mls of a 0.1% trifluoroacetic acid solution,- angiotensin peptides were eluted (2 mls of a 90% acetonitrile, 0.1% trifluoroacetic acid solution, followed by a second elution with 2 mls of 67% methanol, 33% acetonitrile, 0.1% trifluoroacetic) from the column. The eluate was evaporated under nitrogen in a 45°C water bath. Samples were reconstituted in a radioimmunoassay (RIA) buffer (0.1 M K_2HPO_4, 3.0 mM EDTA, 0.15 mM 8-hydroxyquinoline, 0.25% bovine serum albumin). The AII RIA was performed using a polyclonal AII antibody (kindly provided by Dr. Art Freedlender, University of Virginia) which had 1.8% cross reactivity to angiotensin I, and 100% cross reactivity to angiotensin III. The sensitivity of the RIA was 2 pg. The interassay variability was 3.5% and the intraassay variability was 2.0%. Recovery of cold AII using this procedure was 89.7%.

Statistics

Results are represented as mean ± SEM. Statistical analysis was performed using multivariate ANOVA. Duncan's new multiple range test and Dunnet's test were used for post-hoc comparisons.

Results

Preliminary experiments demonstrated a frequency-dependent increase in [3H]-overflow from ISF slices, which was dependent on the presence of extracellular calcium (data not shown). Superfusion with low concentrations (0.1, 1.0 nM) of AII increased (2-fold) stimulation-evoked [3H]-overflow (Fig. 1). However, stimulation-evoked [3H]-overflow in the presence of higher concentrations (10 nM) of AII was not significantly different from control.

To determine if AII is released from ISF, immunoreactive AII release from ISF slices was measured over a 6-h period. Immunoreactive AII release (4.51 ± 0.32 pg/mg tissue wet wt/30 mls collected over a 30 min period) from ISF slices was stable over the 6 hour period (Fig. 2, insert). Immunoreactive AII release from ISF slices was significantly increased (maximum of 2.6-fold) in a delayed manner after S2; however, the release after S1 did not reach significant levels.

To determine if endogenous AII modulates sympathetic neurotransmission, the effect of DuP 753 on evoked [3H]-overflow in ISF slices was examined. In these experiments, [3H]-overflow represents the net result of the events of neuronal release, reuptake, and diffusion of [3H]-metabolites from the presynaptic terminal into the extraceullular space. DuP 753, at concentrations tested, had no effect on basal [3H]-outflow (data not shown). The effect of DuP 753 on the stimulation-evoked [3H]-overflow is illustrated in Fig. 3. DuP 753 in a concentration-dependent (1–100 nM)

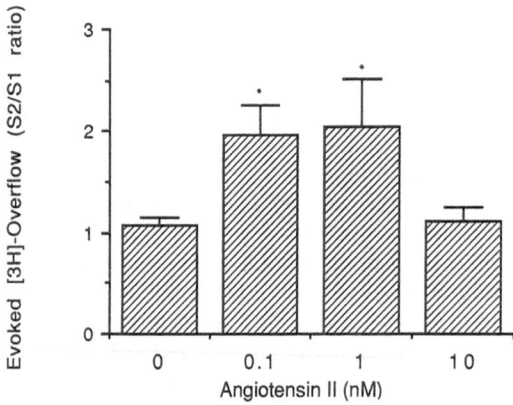

Fig. 1. AII increases the stimulation evoked [3H]-overflow in superfusate from rat ISF slices. Slices were superfused for 90 min prior to S1 (5 Hz for 2 min). AII was added to the superfusion buffer 30 min prior to S2. Data are expressed as mean ±SEM ratio of S2/S1. Basal outflow of [3H] was calculated from the average outflow in two samples collected just prior to S1. The effect of 30 min superfusion with drug on basal outflow was calculated from the average value from 3 samples collected during superfusion with drug. Evoked [3H]-overflow was calculated as the increase in [3H] above basal following electrical stimulation. The AII-induced increase in overflow was significant ($F(3,15) = 4.23$, $P < 0.05$). Asterisks denote significant ($P < 0.05$; N = 6 rats) difference from control (Dunnett's t-test)

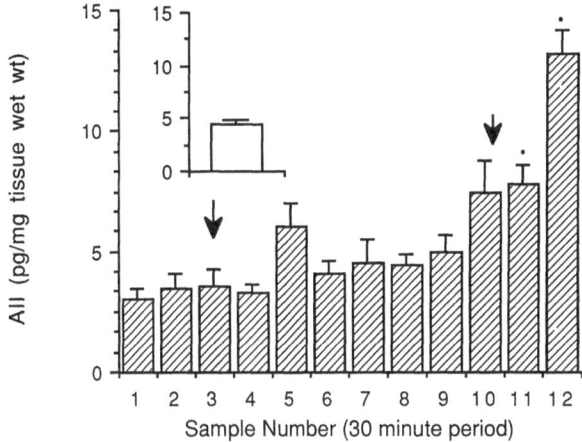

Fig. 2. Stimulation-evoked release of immunoreactive AII from ISF slices over a 6-h period. Insert: Basal immunoreactive AII release from ISF slices was stable and averaged over the 6-h period. Slices were electrically stimulated (10 Hz for 5 min, indicated by arrow) at 60 and 270 min after the start of superfusion. AII immunoreactivity was determined in 30-min samples and is expressed as the mean ±SEM in pg/mg tissue wet wt per 30-min period over 6 h. Electrical stimulation significantly ($F_{(11,33)}$ = 3.31, P < 0.05) increased immunoreactive AII release. Asterisks denote significant (P < 0.05; N = 4 rats) difference from sample #9 (prestimulus) (Duncan's Multiple range test)

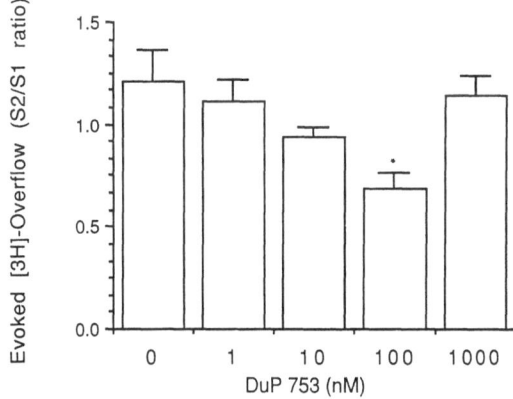

Fig. 3. DuP 753 decreases the stimulation-evoked [3H]-overflow from ISF slices. Experiments were performed as described in Fig. 1. DuP 753 (1 nM–1 μM) was added to the superfusion buffer 30 min prior to S2. Data are expressed as mean ±SEM ratio of S2/S1. DuP 753-induced decrease in evoked [3H]-overflow was analyzed using a repeated measures ANOVA revealing a significant ($F_{(3,15)}$ = 4.01, P < 0.05) main effect of DuP 753. Asterisks denote significant (P < 0.05; N = 6 rats) difference from ISF slices not exposed to DuP 753

manner decreased (maximum of 57%) evoked [3H]-overflow; however, high concentrations (1 μM) of DuP 753 did not significantly alter evoked [3H]-overflow compared to control.

To determine if inhibition of neuronal uptake would alter the DuP 753-induced decrease in evoked [3H]-overflow from ISF slices, the effect of

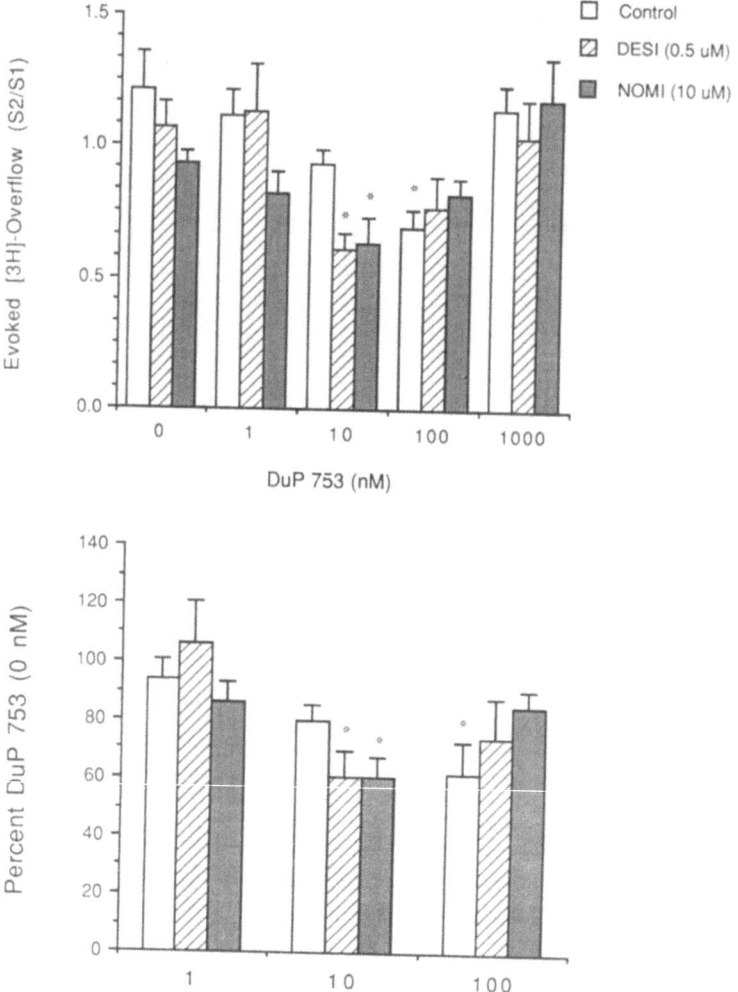

Fig. 4. DuP 753 and evoked [3H]-overflow in the absence and presence of neuronal uptake inhibitors, NOMI or DESI. Experiments were performed as described in Fig. 3; however, NOMI or DESI were present in the superfusion buffer from the start of superfusion. *Top*: Concentration-response curves for DuP 753 on evoked [3H]-overflow in ISF slices in the absence and presence of NOMI or DESI. *Bottom*: Each concentration of DuP 753 was expressed as a percent of response in the absence of DuP 753. Examination of the results using a two factor ANOVA (one between-group factor = exposure to uptake inhibitors, and one within-group factor = DuP 753 concentration) revealed only a significant main effect of DuP 753 concentration (DuP 753 concentration-factor, $F(3,6) = 13.16$, $P < 0.05$; uptake inhibitor-factor, $F(2,17) = 2.11$, $P = 0.1505$; and the interaction term, $F(6,51) = 2.14$, $P = 0.0639$). Dunnett's test was utilized to determine as significant decrease induced by DuP 753 concentration compared to the absence of DuP 753, denoted by the asterisks

neuronal uptake inhibitors, NOMI or DESI, on the DuP 753 concentration-response curve was determined. Initially, concentration-response curves for NOMI and DESI on [3H]-NE uptake into ISF slices were determined, and concentrations (0.5 μM DESI, 10 μM NOMI) were chosen which decreased (maximal 80%) the uptake of [3H]-NE into ISF slices. In the presence of NOMI or DESI, stimulation-evoked [3H]-overflow from ISF slices was significantly increased (2-fold) (data not shown). The concentration-response curve for the inhibitory effects of DuP 753 on evoked [3H]-overflow was such that 10 nM DuP 753 in the presence of uptake inhibitors produced the same inhibitory effect as 100 nM DuP 753 in the absence of uptake inhibitors (Fig. 4, top panel); however, the ANOVA indicated that the interaction between DuP 753 and uptake inhibitors was not significant. Figure 4 (lower panel) illustrates the concentration-response curves for DuP 753 in the absence and presence of uptake inhibitors expressed as a percentage of its contemporaneous control (i.e., absence of DuP 753). Clearly, the addition of neuronal uptake inhibitors did not alter the concentration-response relationship to DuP 753.

Discussion

The present study demonstrates that AII modulates sympathetic neurotransmission in ISF, a source of brown adipose tissue. The use of a 500 μm slice of tissue for examination of release of a neurotransmitter or peptide is a technique which has not been previously applied to ISF. The concentrations of AII in this study which resulted in an increase in evoked [3H]-NE release from ISF slices are in agreement with previous studies examining AII presynaptic effects (Starke, 1977; Westfall, 1977). At the highest concentration (10 nM) of AII examined in this study, evoked [3H]-NE release was no longer significantly different from control, suggesting other AII-mediated responses such as vasoconstriction of blood vessels or increases in prostaglandin synthesis may have predominated.

The basal release of immunoreactive AII from ISF slices did not decrease over 6 h, suggesting de novo synthesis of AII was occurring in ISF. In agreement with these findings, AII tissue levels have been previously determined in ISF (440 pg/g or 8.8 pg/20 mg slice; Phillips et al., 1991). In this study, basal immunoreactive AII release was 4.5 pg/mg tissue wet wt/30 min which corresponds to release of 3 pg AII/20 mg ISF slice per minute, or 34% of ISF AII levels. Electrical field stimulation of ISF slices resulted in a delayed increase in superfusate concentrations of immunoreactive AII. Interestingly, immunoreactive AII concentrations significantly increased only after a second stimulus was applied to ISF slices, indicating the necessity for a priming stimulus for AII release. Considering that electrical field stimulation resulted in the immediate release of [3H]-NE from ISF (data not shown), the delayed increase in AII release may be related to release of enzymes involved in the synthesis of AII, rather than release of AII stored in nerve endings. The cell type(s) involved in synthesis and the enzymes responsible for AII production in brown adipose tissue are not known, and this will be the focus of future experiments.

Basal and evoked immunoreactive AII release from ISF slices in this study resulted in extracellular AII concentrations of 15 and 43 pM, respectively. Experiments examining the effect of exogenous AII on evoked [3H]-

NE overflow from ISF slices demonstrated a significant increase in evoked [3H]-overflow in response to 100 pM AII; however, lower concentrations of AII were not examined. The nonpeptide specific AII-1 receptor antagonist (Chiu et al., 1990; Wong et al., 1990), DuP 753, resulted in a decrease in evoked [3H]-NE release from ISF slices. Thus, as evidenced by the effect of DuP 753, the endogenous AII concentration in ISF was sufficient to enhance sympathetic neurotransmission.

Previous studies have demonstrated that AII has the ability to inhibit neuronal uptake of NE in certain vascular beds (Khairallah, 1972). In the present study, NOMI and DESI significantly increased the evoked release of [3H]-NE. Thus, in ISF, the process of neuronal uptake is important in removal of NE from the synapse. However, NOMI and DESI did not alter the DuP 753 concentration-response curve for evoked release of [3H]-NE. These results suggest that the facilitation of sympathetic neurotransmission in ISF in response to AII is most likely due to enhanced release of NE, rather than to inhibition of neuronal uptake.

In summary, the present study demonstrates the release of immunoreactive AII from ISF slices, and the ability of exogenous and endogenous AII to modulate sympathetic neurotransmission in brown adipose tissue. Our previous studies have demonstrated the presence of Ao mRNA in ISF. This study suggests that angiotensinogen is processed to AII in ISF, which serves to regulate sympathetic neurotransmission. Therefore, locally produced AII may have a physiological role in regulation of thermogenesis, the major function of brown adipose tissue.

Acknowledgements

We thank D. Painter for her excellent technical assistance. The authors are also grateful for the kind gifts of DuP 753 (E. I. Dupont de Nemours, DE) and Nomifensine (Hoechst-Roussel Pharmaceuticals, Inc, NJ). This work was supported by NIH LBI #R29 HL41954 and a University of Kentucky Medical Center Research Grant.

References

Cannon B, Nedergaard J (1985) The biochemistry of an inefficient tissue: brown adipose tissue. Essays Biochem 20: 110–164
Cassis L, Lynch K, Peach M (1988a) Localization of angiotensinogen messenger RNA in rat aorta. Circ Res 62: 1259–1262
Cassis L, Saye J, Peach M (1988b) Location and regulation of rat angiotensinogen messenger RNA. Hypertension 11: 591–596
Chiu A, McCall D, Price W, Wong P, Carini D, Duncia J, Wexler R, Yoo S, Johnson A, Timmermars P (1990) Nonpeptide angiotensin II receptor antagonists. VII. Cellular and biochemical pharmacology of DuP 753, an orally active antihypertensive agent. J Pharmacol Exp Ther 252: 711–718
Cottle W, Cottle W (1969) Adrenergic fibers in brown fat. Proc Can Fed Biol Soc 12: 994

Cottle W, Cottle W, Perusset F, Bukowiecki L (1985) An improved glyoxylic acid technique for the histochemical localization of catcholamines in brown adipose tissue. Histochem J 17: 1279–1288

Derry D, Schonbaum E, Steiner G (1969) Two sympathetic nerve supplies to brown adipose tissue of the rat. Can J Physiol Pharmacol 47: 57–63

Dwoskin L. Zahniser N (1986) Robust modulation of [^{3}H] dopamine release from rat striatal slices by D-2 dopamine receptors. J Pharmacol Exp Ther 239(2): 442–453

Foster DO, Frydman ML (1978) Non-shivering thermogenesis in the rat. II. Measurements of blood flow with microspheres point to brown adipose tissue as the dominant site of calorigenesis induced by noradrenaline. Can J Physiol Pharmacol 58: 915–924

Foster D, Depocas F, Zuker M (1982) Heterogeneity of the sympathetic innervation of rat interscapular brown adipose tissue via intercostal nerves. Can J Physiol Pharmacol 60: 747–754

Khairallah P (1972) Action of angiotensin on adrenergic nerve endings: inhibition of norepinephrine uptake. Fed Proc 31(4): 1351–1357

Lever J, Norman D, Symons D, Jung R (1986) Evidence of peptidergic nerves within rat brown adipose tissue. J Anat 149: 230

Menard J, Bouhnik J, Clauser E, Richoux J, Corvol P (1983) Biochemistry and regulation of angiotensinogen. Clin Exp Hypertens (A) 5: 1005–1019

Norman D, Mucherjee S, Symons D, Jorg R, Lever J (1988) Neuropeptides in interscapular and perirenal brown adipose tissue in the rat: a plurality of innervation. J Neurocytol 17: 305–311

Phillips M, Kimura B, Baizada M (1991) Angiotensin and atrial natriuretic peptide in tissue and cell culture. Methods Neurosci 4: 177–206

Starke K (1977) Regulation of noradrenaline release by presynaptic receptor systems. Rev Physiol Biochem Pharmacol 77: 1–124

Westfall TC (1977) Local regulation of adrenergic neurotransmission. Physiol Rev 57: 659–728

Wong P, Price W, Chiu A, Duncia J, Carini D, Wexler R, Johnson A, Timmermans P (1990) Nonpeptide angiotensin II receptor antagonists. VIII. Characterization of functional antagonism displayed by DuP 753, an orally active hypertensive agent. J Pharmacol Exp Ther 252: 719–725

Authors' address: L. A. Cassis, Ph.D., Division of Pharmacology and Experimental Therapeutics, Rose Street, College of Pharmacy, University of Kentucky, Lexington, KY 40536-0082, U.S.A.

J Neural Transm (1991) [Suppl] 34: 139–145

Selective destruction of preganglionic sympathetic nerves by antibodies to acetylcholinesterase

S. Brimijoin[1] and **V. A. Lennon**[2]

Departments of [1] Pharmacology and [2] Immunology and Neurology, Mayo Clinic,
Rochester, MN, USA

Summary. Systemic injection of monoclonal antibodies to neural acetyl-cholinesterase in rats causes permanent, complement-mediated destruction of presynaptic fibers in sympathetic ganglia and adrenal medulla. Ptosis, hypotension, bradycardia, and postural syncope ensue. In sympathetic ganglia, cholinergic synapses disappear, but postganglionic adrenergic neurones remain structurally and functionally normal. Somatic motor and parasympathetic systems are also spared. This model of selective cholinergic autoimmunity is a new tool for autonomic physiology and may be relevant to the pathogenesis of human dysautonomias.

Introduction

We are studying the immunopathology of acetylcholinesterase (AChE), an enzyme of interest for its role in neurotransmission, its catalytic power, its structural polymorphism, and its involvement in neurologic disease (Toutant and Massoulié, 1988; Brimijoin and Rakonczay, 1986; Rakonczay and Brimijoin, 1988). AChE is enriched in cholinergic synapses but occurs elsewhere also. At the synapse there are different molecular forms of AChE, divided into 1) simple aggregates of globular catalytic subunits and 2) asymmetric assemblies with a collagen-like tail. The major globular forms of mammalian skeletal muscle and nervous tissue are monomers (mostly intracellular) and tetramers (partially exposed on the surface membranes of muscle and neurones). Asymmetric forms are primarily found at the neuromuscular junction; one form with a sedimentation coefficient of 16S is a specific marker of the rat motor endplate.

AChE is affected directly or indirectly in disorders involving degeneration of cholinergic neurones, such as Alzheimer's disease (Atack et al., 1983; Fishman et al., 1986; Hammond and Brimijoin, 1988), dystrophies and other abnormalities of skeletal muscle (Skau, 1983). Fortunately, cholinergic neurotransmission has a large margin of safety. At the neuromuscular junction, over 75% of the AChE must be inactivated before transmission fails (Koelle, 1963). Greater enzyme loss has drastic consequences. At the motor endplate these include depolarization blockade of synaptic transmission, paralysis and death. At autonomic effector organs, a deficit of AChE leads to parasympathetic hyperstimulation, with intense salivation, tracheal secretion, diarrhea, vomiting, bradycardia, and hypotension. This syndrome can also be fatal.

Because of AChE's pathophysiological importance, we posed the questions: (1) Is AChE vulnerable to immunologic attack, given its proximity to nicotinic acetylcholine receptors (nAChR) and presynaptic calcium channels, the antigenic targets in myasthenia gravis and Lambert-Eaton syndrome? (2) What would be the consequences of autoimmunity to AChE? With these issues in mind we constructed an animal model of passive AChE-autoimmunity based upon monoclonal antibodies to rat brain AChE. The present paper provides a review of published findings (Brimijoin et al., 1990; Brimijoin and Lennon, 1990), and a summary of current results.

Results

Distribution of antibody and access to antigen

Monoclonal IgG antibodies (MAbs) to rat brain AChE (ZR series, Rakonczay and Brimijoin, 1986) may be injected into the tail veins of adult rats to produce passive autoimmunity. None of our antibodies directly affects AChE's catalytic activity. Therefore, changes in measured enzyme activity reflect changes in the amount of enzyme present. Much of the murine IgG remains in the blood circulation for days, but biologically significant amounts reach antigenic targets in muscle and sympathetic ganglia. In ganglia, one third of the AChE is complexed with IgG. In muscle, the figure is higher and there is impressive complexation at the motor endplate. Sucrose density gradient analyses of muscle extracts prepared 3 days after injection of AChE-antibody show that *all* of the synapse-specific 16S AChE is complexed with IgG. In contrast, 4S AChE, an intracellular enzyme form, remains free of antibody. The antibodies used in these experiments are effectively excluded by the blood brain barrier and do not accumulate in the CNS.

Effects of systemically administered AChE-antibodies

It was discovered that AChE-antibodies trigger a devastating immunological attack on preganglionic sympathetic neurones, with minimal effects on other cholinergic systems. A large body of evidence supporting this view is summarized in Table 1. Some details are given below.

Rats injected i.v. with AChE antibodies (1.5 mg of IgG, equal mixture of monoclonals ZR 2, 3, 4, 5, & 6) develop a profound sympathetic disturbance without obvious dysfunction in the motor or parasympathetic nervous

Table 1. Peripheral effects of AChE antibodies in adult rats

	Preganglionic sympathetic system	Postganglionic sympathetic system	Parasympathetic system	Somatic motor system
Clinical signs	Rapid onset of permanent eyelid drooping (ptosis). Syncope	No specific signs	No excess salivation or diarrhea	Normal grip strength, righting reflex, locomotion
Physiology	No eyelid response to stimulation of cervical sympathetic chain. Persistent hypotension and bradycardia. Impaired stress-induced catecholamine release	Normal eyelid response to direct electrical stimulation of SCG	Slowed heart beat on vagal stimulation, normal frequency-response	not tested
Pharmacology	not tested	Normal pressor responses to noradrenaline, tyramine, and ganglionic agonists (DMPP)	Normal chronotropic dose-response curve for ACh on isolated heart	not tested
Biochemistry	Minor loss of AChE and total loss of ChAT activity in sympathetic ganglia and adrenal glands	Normal DBH activity in superior cervical ganglion	Normal cardiac ChAT activity	Normal AChR and ChAT in diaphragm. Selective loss of asymmetric and tetrameric globular AChE forms
Structure	AChE staining disappears from neuropil of SCG. Intense punctate deposits of IgG and complement mark presumptive synapses	Principal neurons in SCG show normal Nissl and AChE staining. Tyrosine hydroxylase immunocyto-chemistry is normal	Intramural ganglia of atria show near normal AChE staining but contain IgG and complement deposits by immunocyto-chemistry	Diaphragm endplates normal in number and appearance by light microscopy. IgG and complement deposition evident by immunocyto-chemistry
Ultrastructure	Vesicle-laden synaptic terminals disappear from SCG. Dendrites and non-terminal axons appear normal. Lipid deposits accumulate within Schwann cells	Endoplasmic reticulum, mitochondria and other organelles of principal neurons in SCG appear structurally intact	not tested	Normal folding pattern at motor endplate. No evidence of damage to motor neuron

Abbreviations: *ACh* acetylcholine, *AChE* acetylcholinesterase, *AChR* acetylcholine receptors, *ChAT* choline acetyltransferase, *DBH* dopamine-beta-hydroxylase, *DMPP* dimethylphenyl-piperazinium, *IgG* immunoglobulin G, *SCG* superior cervical ganglion systems. Over 60 rats have been observed for periods up to several months. Injection of normal mouse IgG has no apparent effect. But all rats receiving AChE antibodies develop signs of transient sympathetic activation followed by prolonged sympathetic depression. They remain clinically normal otherwise. We do not see muscle fasciculation or obvious motor dysfunction. Moreover, unlike anticholinesterase drugs, AChE-antibodies cause no parasympathetic overactivity (lacrimation, salivation, or diarrhea).

Sympathetic abnormalities appear quickly. Piloerection and exophthalmos begin 20 min after injection, then fade. Next comes eyelid drooping (ptosis, a sign of sympathetic denervation). Ptosis is maximal by 2 h, and continues indefinitely. A rat observed for 15 months had ptosis until death, although serum antibody was not detectable after 2 months. Because AChE stores in nerve and muscle are replaced in a few weeks, persistent ptosis must reflect permanent cellular lesions, not mere enzyme loss.

Tonic control of eyelid smooth muscle is adrenergic (Isola and Bacq, 1946). Thus a likely site for a ptosis-inducing lesion is the superior cervical ganglion. Damage to the ganglion was first demonstrated by recording palpebral tension while electrically stimulating the presynaptic, cervical sympathetic chain. In control rats eyelid tension increased with preganglionic and direct ganglionic stimulation. Eyelid contraction was also induced by dimethylphenylpiperazinium (DMPP), a selective agonist for nAChR of ganglia. Rats with ptosis, tested 3 to 90 days after injection of AChE-antibodies, failed to respond to preganglionic stimulation, but all responded normally to direct ganglionic stimulation and to DMPP. The postganglionic neurones were intact but the preganglionic terminals must have been disabled.

Another striking sign of autonomic disturbance, never seen in controls, is postural syncope. Starting 6 h after injection, treated rats lose consciousness and stop breathing when gripped tightly about the thorax and held upright. They recover quickly when placed on their backs. These observations suggest that sympathetic damage affects cardiovascular control mechanisms.

Heart rate and mean blood pressure are acutely disturbed by AChE-antibody. Typically, pressure and rate are stable for the first 20 min after antibody-injection, then there is transient hypertension and cardioacceleration. By 2 h heart rate and blood pressure fall below normal and remain depressed for weeks. Atropine does not raise blood pressure or heart rate at this stage. Hence, the cardiovascular effects of antibody do not reflect buildup of acetylcholine at muscarinic receptors.

Damage to adrenergic neurones has been ruled out by another pharmacological experiment. In control rats, blood pressure rises in response to the ganglionic agonist, DMPP. DMPP also raises blood pressure in hypotensive rats injected with AChE-antibodies. Evidently, the postganglionic portion of the sympathoadrenal pressor system, including the nAChR of ganglia, is fully functional. Parasympathetic function is also intact, as shown by a normal negative chronotropic response to vagal stimulation.

Enzyme assays and morphology have confirmed the selective damage of preganglionic sympathetic neurones. At every time tested, the activity of dopamine-β-hydroxylase in the superior cervical ganglia has been 80% of normal. Therefore, adrenergic neurones, for which this enzyme is a marker, are largely spared. Histochemical staining at 1 to 90 days after antibody-injection reveals striking and selective loss of AChE activity in ganglionic neuropil but not in nerve cell bodies.

In electron micrographs, superior cervical ganglia of control rats have many vesicle-laden terminals juxtaposed to pre- and postsynaptic densities. Ganglia examined one week after AChE-antibody have no bonafide synapses. Loss of synaptic structures is paralleled by loss of choline acetyltransferase activity (ChAT), a marker of cholinergic cytoplasm. ChAT activity *vanishes* from the superior cervical ganglion 3 days after injection. It is also depleted in stellate, thoracic, lumbar and coeliac ganglia, and adrenal gland, implying widespread destruction of presynaptic terminals. On the other hand, normal ChAT activity in diaphragm and atria indicates survival of vagal terminals and ganglia.

Experiments with purified anti-complementary factor from *Naja naja* cobra venom clarified the mechanism of immunosympathectomy. In rats depleted of hemolytic complement by cobra venom factor, antibody-injection did not cause ptosis, and the sympathetic ganglia did not lose ChAT activity. Recently we have observed ganglionic accumulations of IgG and complement (C3) within a few hours of antibody-injection. Complement activation is obviously needed for the antibody-induced lesions.

Discussion and conclusions

Our results show that AChE antibodies induce a permanent, global, and selective destruction of preganglionic sympathetic nerve terminals. This pattern was unexpected since the neuromuscular junction and motor neurone are also very rich in the target enzyme and are known to be vulnerable to immunologic damage. The unique syndrome of preganglionic sympathectomy differs from every previously described form of sympathectomy and is not induced by other antibodies or toxins. The persistence and selectivity of autoimmune damage by AChE antibodies opens up interesting new lines of investigation as follows.

Persistance of the preganglionic lesion

Permanent ptosis implies that preganglionic terminals in the superior cervical ganglion fail to regenerate after lesioning by AChE-antibodies. This failure contrasts with the robust regeneration that follows surgical lesions of the cervical sympathetic chain (Raisman et al., 1974). Preliminary results suggest that presynaptic neurones survive for several days but may die between 1 and 4 months after antibody-injection. It remains to be determined if the poor recovery from antibody-mediated damage primarily reflects delayed death of the presynaptic neurone or failure to regenerate axonal processes despite neuronal survival.

Locus of immunologic damage

There are at least three potential explanations for selective antibody-induced damage of preganglionic sympathetic neurones with sparing of motor neurones: (1) IgG might have better access to ganglionic synapses than to the neuromuscular junction; (2) the abundance of extracellular AChE in the basal lamina of the motor endplate might shift the focus of immunologic attack away from the neuronal plasma membrane; (3) the molecular architecture or topology of ganglionic AChE might favor complement-mediated damage. At present we cannot exclude any of these explanations, but a protective role for AChE in the basal lamina seems most likely. The enriched assembly of multisubunit forms of enzyme in this location represents a set of densely packed, repeating epitopes. This antigenic "sink" for IgG and for activation of complement could limit immunologic damage of adjacent cells.

Applications and implications

The rat model of AChE-autoimmunity lends itself to studies of autonomic physiology and pathophysiology. A unique feature of preganglionic immunosympathectomy is the potential of MAbs to block all impulse-related release of norepinephrine and epinephrine, without any effects on vasomotor centers in the brain. This feature contrasts with guanethidine-induced sympathectomy (Burnstock et al., 1971; Johnson and O'Brien, 1976) or immunosympathectomy with antibodies to nerve growth factor (Levi Montalcini and Angeletti, 1966), both of which target *postganglionic* neurones and spare the adrenal gland. Our model should be especially useful for (1) investigating the role of the peripheral sympathoadrenal system in spontaneously hypertensive rats, (2) defining the peripheral actions of antihypertensive drugs, or (3) studying the activity-dependent regulation of adrenergic receptors in the vasculature. Furthermore, from a clinical standpoint, the pathophysiology of experimental AChE autoimmunity is sufficiently similar to certain human disorders to warrant critical study of the underlying mechanisms.

Acknowledgements

This work was supported by grants from the National Institute of Neurological and Communicative Disorders and Stroke (NS 181790, NS 29646 and NS 15057). We thank P. Hammond for technical assistance.

References

Atack JR, Perry EK, Bonham JR, Perry RH, Tomlinson BE, Candy J, Blessed G, Fairbairn A (1983) Molecular forms of acetylcholinesterase in senile dementia of Alzheimer type: selective loss of the intermediate (10S) form. Neurosci Lett 40: 199–204

Brimijoin S, Rakonczay Z (1986) Immunology and molecular biology of the cholinesterases. Int Rev Neurobiol 28: 363–410

Brimijoin S, Lennon VA (1990) Autoimmune preganglionic sympathectomy induced by acetylcholinesterase antibodies. PNAS 87: 9630

Brimijoin S, Balm M, Hammond P, Lennon VA (1990) Selective complexing of acetylcholinesterase in brain by intravenously administered monoclonal antibody. J Neurochem 54: 236–241

Burnstock G, Evans B, Gannon BJ, Heath JW, James V (1971) A new method of destroying adrenergic nerves in adult animals using guanethidine. Br J Pharmacol 43: 295–301

Fishman EB, Siek GC, MacCallum RD, Bird ED, Volicer L, Marquis JK (1986) Distribution of the molecular forms of acetylcholinesterase in human brain: alterations in dementia of the Alzheimer type. Ann Neurol 19: 246–252

Hammond P, Brimijoin S (1988) Acetylcholinesterase in Huntington's and Alzheimer's diseases: simultaneous enzyme assay and immunoassay of multiple brain regions. J Neurochem 50: 1111–1116

Isola W, Bacq ZM (1946) Innervation sympathique adrénergique de la musculature lisse des paupières. Arch Int Physiol 54: 30–48

Johnson EM, O'Brien F (1976) Evaluation of the permanent sympathectomy produced by the administration of guanethidine to adult rats. J Pharmacol Exp Ther 196: 53–61

Koelle GB (1963) Cytological distributions and physiological functions of cholinesterases. In: Koelle GB (ed) Cholinesterases and anticholinesterase agent. Springer, Berlin Heidelberg, pp 187–298

Levi-Montalcini R, Angeletti R (1966) Immunosympathectomy. Pharmacol Rev 18: 619–628

Raisman G, Field PM, Ostberg AJC, Iversen LL, Zigmond RE (1974) A quantitative ultrastructural and biochemical analysis of the process of reinnervation of the superior cervical ganglion in the adult rat. Brain Res 71: 1–16

Rakonczay Z, Brimijoin S (1986) Monoclonal antibodies to rat brain acetylcholinesterase: comparative affinity for soluble and membrane-associated enzyme and for enzyme from different vertebrate species. J Neurochem 46: 280–287

Rakonczay Z, Brimijoin S (1988) Biochemistry and pathophysiology of the molecular forms of cholinesterases. Subcell Biochem 12: 335–378

Skau KA (1983) The acetylcholinesterase abnormality in dystrophic mice is a reflection of a maturational defect. Brain Res 276: 192–194

Toutant J-P, Massoulié J (1988) Cholinesterase: tissue and cellular distribution of molecular forms and their physiological regulation. Handbook Exp Pharmacol 86: 225–265

Authors' address: S. Brimijoin, Ph.D., Department of Pharmacology, Mayo Medical School, Rochester, MN 55905, U.S.A.

References

Veen, W., ...
...

Receptors and post-receptor events

J Neural Transm (1991) [Suppl] 34: 149–155

Adenosine receptors in the central nervous system

U. Schwabe, A. Lorenzen, and **S. Grün**

Department of Pharmacology, University of Heidelberg, Federal Republic of Germany

Summary. Two major subclasses of adenosine receptors have been distinguished in the central nervous system, termed A_1 and A_2. They are coupled to G-proteins and regulate the activity of adenylyl cyclase, potassium channels and several other effector systems. Autoradiographic studies have shown that A_1 receptors are mainly found in the hippocampus and the cerebellum, whereas A_2 receptors are almost exclusively located in the striatum and olfactory tubercle. Furthermore, a novel adenosine binding protein was identified in bovine striatum by radioligand binding with [^{3}H]5′-N-ethylcarboxamidoadenosine ([^{3}H]NECA). The pharmacological profile of this NECA binding protein has been determined in competition experiments with adenosine receptor ligands. It can be distinguished from that of A_2 adenosine receptors and other adenosine binding proteins such as S-adenosylhomocysteine hydrolase and the adenosine transporter.

Introduction

Adenosine has long been recognized as an important neuromodulator with profound effects on the activity of the central nervous system (Williams, 1989). It has well-documented inhibitory effects on the release of a number of neurotransmitters such as acetylcholine, noradrenaline, dopamine, serotonin, glutamate and aspartate. These effects occur via presynaptic adenosine receptors. Furthermore, adenosine and metabolically stable adenosine analogs have marked sedative and anticonvulsant properties. Adenosine is also able to modulate antinociceptive effects and to depress spontaneous locomotor activity independent from its peripheral effects on the cardiovascular system.

These findings have led to the idea that adenosine mediates a general "inhibitory tone" in the central nervous system and thus might have a specific protective function under conditions of enhanced neuronal activity which always are associated with an increased release of adenosine. These responses appear to be mediated in most cases by highly specific adenosine receptors that have been divided into at least two major subclasses termed A_1 and A_2. Both adenosine receptor subtypes were originally defined on the basis of adenylyl cyclase studies (Londos et al., 1980).

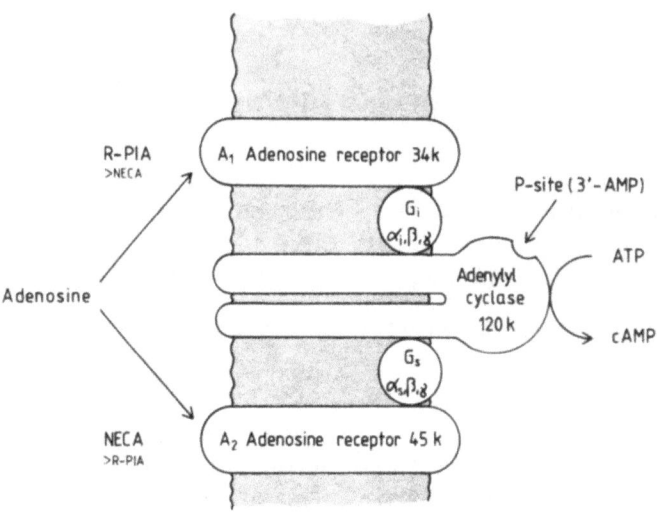

Fig. 1. Adenosine receptors coupled to adenylyl cyclase

A_1 and A_2 receptors are linked to different components of the adenylyl cyclase system. The A_1 receptor interacts with G_i and inhibits cyclase while the A_2 receptor interacts with G_s and activates the enzyme (Fig. 1). Both adenosine receptors have been characterized by additional pharmacological criteria and can be discriminated by the two agonists $R-N^6$-phenylisopropyladenosine (R-PIA) and 5'-N-ethylcarboxamidoadenosine (NECA). R-PIA is more potent than NECA at the A_1 receptor while the reverse potency order is observed at the A_2 receptor.

Furthermore, photoaffinity ligands were developed from the derivatives of R-PIA and NECA which can be covalently incorporated into the binding unit of both receptor subtypes. With this approach it has been shown that the A_1 receptor has an apparent molecular weight of approximately 34 kDa and the A_2 receptor of 45 kDa (Klotz et al., 1985; Barrington et al., 1989). A second approach to elucidate the structure of adenosine receptors involves the purification of the receptor protein by affinity chromatography. With this procedure the A_1 receptors from rat and bovine brain have been purified to apparent homogeneity (Nakata, 1989; Munshi and Linden, 1989). More recently, the structure of both receptor subtypes has been elucidated by cloning the receptors from a dog thyroid cDNA library (Maenhaut et al., 1990; Libert et al., 1991).

In addition to adenylyl cyclase several other effector systems have been described which are modulated by A_1 receptors in a positive or negative manner. The best characterized effector system is the potassium conductance in atrial cells and hippocampal neurons. Probably it is the same potassium channel which is activated by muscarinic receptors, the so-called receptor-coupled potassium channel. Adenosine also inhibits voltage-sensitive calcium channels in hippocampus and other neuronal tissues. Probably it is the N-type channel which appears to be responsible for the calcium influx triggering transmitter release. Further effector systems appear to be coupled to the A_1 receptor such as phospholipase C, guanylyl cyclase and cyclic AMP phosphodiesterase. As already mentioned, the A_2 adenosine receptor is only coupled to adenylyl cyclase for instance in striatum and human platelets.

Autoradiographic studies on brain adenosine receptors

Several adenosine receptor agonists and antagonists have been used for the characterization of the two receptor subtypes in brain tissue (Fig. 2). R-PIA was originally used for the pharmacological definition of A_1 receptors in cyclase studies (Londos et al., 1980). It is still considered as a kind of standard agonist for this receptor subtype. The A_1 affinity is 1.3 nM and the A_1 selectivity as compared to A_2 receptors is 560-fold. In the search for more potent and more selective agonists many adenosine derivatives have been synthesized during the last 10 years. The most suitable compound appears to be the 2-chloro-N^6-cyclopentyladenosine (CCPA) which was developed from R-PIA by two minor modifications (Lohse et al., 1988). It is 3-fold more potent than R-PIA and has almost a 10,000-fold A_1-selectivity.

1,3-Dipropyl-8-cyclopentylxanthine (DPCPX) is an adenosine receptor antagonist and was developed from the well-known methylxanthines theophylline and caffeine. DPCPX is much more potent than theophylline and has a more than 1,000-fold selectivity for the A_1 receptor (Lohse et al., 1987). At present, it is the most suitable A_1 antagonist for functional and radioligand binding studies.

We have used this compound to study the autoradiographic localization of adenosine receptors in brain slices (Weber et al., 1990). The distribution of specific [^{3}H]DPCPX binding to a rat brain sagittal section was examined at a radioligand concentration of 0.8 nM. Regions of very high receptor density were the hippocampus, the gyrus dentatus and the molecular layer of cerebellar cortex. Most hypothalamic nuclei and certain layers of the cerebral cortex also exhibited high levels of binding whereas other cortical layers and the striatum showed only moderate labeling.

Compared to autoradiographic studies with agonist radioligands the brain structures with high A_1 adenosine receptor densities are visualized much more clearly, since antagonist radioligands label both the high and the low affinity state of adenosine receptors.

In the A_2 receptor field mainly agonists have been used for the characterization of this receptor subtype (Fig. 3). NECA has already been mentioned as the prototypic agonist for A_2 receptors. However, NECA has a higher affinity for A_1 receptors in many tissues and therefore is not A_2-selective (Bruns et al., 1986). The search for more selective agonists was at least partially successful. The most interesting compound is 2-[p-(2-carboxyethyl)-phenethylamino]-5'-N-ethylcarboxamidoadenosine (CGS 21680), a NECA-derivative with a large substituent at the 2-position of the purine moiety. This compound has a considerably improved selectivity for A_2 receptors (Hutchison et al., 1989). This is mainly due to the low affinity

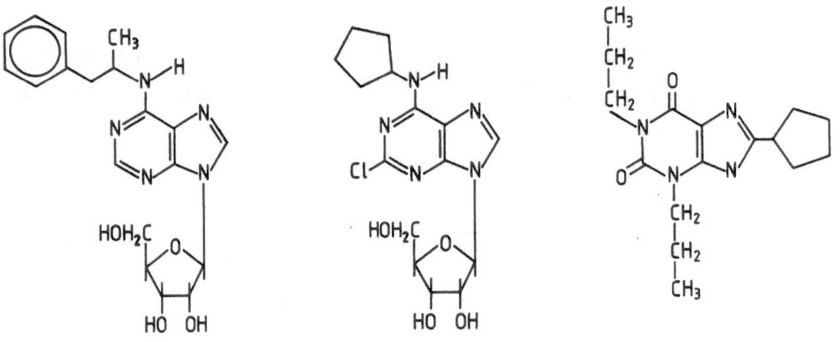

R–N[6]-Phenylisopropyladenosine   2-Chloro-N[6]-cyclopentyladenosine   1,3-Dipropyl-8-cyclopentylxanthine
R–PIA   CCPA   DPCPX

A_1 Affinity (K_i)	1.3 nM	0.4 nM	0,3 nM
A_1 Selectivity	560	9,750	1,130

Fig. 2. Structures of A_1 adenosine receptor ligands

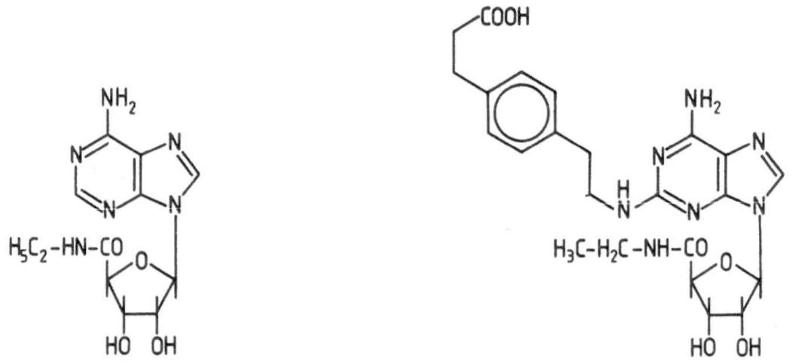

5'–N-Ethylcarboxamido-   CGS 21680
adenosine, NECA   2-[p-(2-carboxyethyl)phenethylamino-
                                5'-N-ethylcarboxamidoadenosine

A_2 Affinity (K_i)	12 nM	22 nM
A_2 Selectivity	0.5	140

Fig. 3. Structures of A_2 adenosine receptor agonists

at A_1 receptors whereas the A_2 affinity is only 22 nM and thus even lower than that of NECA.

Radiolabeled CGS 21680 has also been used to study the autoradiographic distribution of A_2 receptors in rat brain. The binding of [³H]CGS 21680 to a sagittal rat brain section was studied at a concentration of 5 nM. Analysis of the specific binding showed that A_2 adenosine receptors are only associated with the caudate putamen, nucleus accumbens and olfactory tubercle, whereas no binding was detected elsewhere (Jarvis and Williams, 1989). These results demonstrate a specific localization of A_2 receptors in the striatal region of rat brain and indicate a specific role of A_2 adenosine receptors in basal ganglia.

Characterization of a novel NECA binding protein

In order to achieve a more detailed characterization of the striatal A_2 receptor we have studied radioligand binding in bovine striatum after solubilization and gel filtration of the receptors (Lorenzen et al., 1991). The typical elution profile after gel filtration on Sepharose CL-6B yielded three [^{3}H]NECA binding peaks. The first peak was eluted with the void volume of the column together with the major protein peak and contained 10–20% of the total NECA binding. The second and the third peak were clearly separated and contained the majority of the [^{3}H]NECA binding. If non-specific binding was defined with $100\,\mu M$ 2-chloroadenosine, [^{3}H]NECA was displaced in all three peaks. In contrast $100\,\mu M$ R-PIA displaced [^{3}H]NECA only from the first and the third peak but not from the second peak. These binding properties might indicate that peak II contains NECA binding sites resembling a NECA binding protein which was previously identified in human platelets and human placenta with a high degree of homology to certain stress proteins. However, peak III appeared to be different from the two other [^{3}H]NECA binding proteins.

Therefore, we have studied the saturation of these binding proteins. Specific [^{3}H]NECA binding to peak III from bovine striatal membranes was saturable in a single component as shown by the Scatchard analysis of the binding. The B_{max} value for the binding capacity was $11.3\,pmol/mg$ protein and the K_D value was $16.9\,nM$. Thus, peak III had a very high affinity for [^{3}H]NECA which is equivalent to the affinity of the A_2 receptor for NECA.

We have compared the saturation data of peak I and peak III for [^{3}H]NECA and the A_2-selective radioligand [^{3}H]CGS 21680. The affinity of [^{3}H]NECA to peak I and peak III was almost identical as shown by K_D values of 21 and $17\,nM$. Similar binding data were obtained with the CGS compound in peak I whereas no specific binding of CGS was obtained in peak III fractions. These data indicate that peak I might be equivalent to the A_2 adenosine receptor whereas peak III might be a different NECA binding protein.

For a more detailed characterization of both binding peaks we have studied the competition of [^{3}H]NECA by different adenosine receptor ligands (Table 1). The K_i values of the first peak agree well with their affinities for the striatal A_2 receptors in membrane preparations. Particularly high affinities were obtained for NECA, CGS 21680 and the adenosine antagonist xanthine amine congener (XAC). In contrast, most of the adenosine receptor ligands did not compete significantly for [^{3}H]NECA binding to peak III. High affinities were only observed for NECA itself and inosine, which is inactive at the A_2 and also the A_1 receptor. Typical A_1 receptor compounds such as 2-chloro-N^6-cyclopentyladenosine (CCPA) and XAC were also inactive. Furthermore, a very low affinity was observed for the P_{2y} agonist 2-methylthio-ATP, indicating that we have not isolated one of the well-known ATP receptor subtypes.

We have further excluded another well-known adenosine binding protein, the adenosine transporter, since the adenosine uptake inhibitor

Table 1. Pharmacological profile of $[^3H]$ NECA binding sites from bovine striatum

Compound	A_2 receptor (peak I) K_i nmol/l	Peak III K_i nmol/l
NECA	20	23
CGS 21680*	19	>100,000
CCPA	1,660	87,000
Inosine	>100,000	71
XAC	20	102,000
Theophylline	22,000	3,900,000
2-Methylthio-ATP	n.d.	97,000
Dipyridamole	n.d.	>10,000
Adenosine-2′,3′-dialdehyde	n.d.	25,000

*2-p-Carboxyethyl phenylamino-5′-N-ethylcarboxamidoadenosine; *n.d.* not determined

dipyridamole was inactive at concentrations up to 10 μM. Finally, the S-adenosylhomocysteine hydrolase (SAH-hydrolase) was excluded, since the highly selective inhibitor adenosine-2′,3′-dialdehyde was only effective at 25 μM, whereas the SAH-hydrolase is blocked irreversibly already at low nanomolar concentrations.

These data show that the NECA binding protein in peak III can be distinguished from the G-protein-coupled adenosine receptors and several other adenosine binding proteins. At present, we know that this protein is located in membrane fractions but we have no information on the subcellular distribution. The nature and the function of this additional NECA binding protein remain to be elucidated. We hope that the purification and the analysis of the structure of this protein will provide further information on the function and a possible physiological role.

References

Barrington WW, Jacobson KA, Hutchison AJ, Williams M, Stiles GL (1989) Identification of the A_2 adenosine receptor binding subunit by photoaffinity crosslinking. Proc Natl Acad Sci USA 86: 6572–6576

Bruns RF, Lu GH, Pugsley TA (1986) Characterization of the A_2 adenosine receptor labeled by $[^3H]$NECA in rat striatal membranes. Mol Pharmacol 29: 331–346

Hutchison AJ, Webb RL, Oei HH, Ghai GR, Zimmerman MB, Williams M (1989) CGS 21680C, an A_2 selective adenosine receptor agonist with preferential hypotensive activity. J Pharmacol Exp Ther 251: 47–55

Jarvis MF, Williams M (1989) Direct autoradiographic localization of adenosine A_2 receptors in the rat brain using the A_2-selective agonist, $[^3H]$CGS 21680. Eur J Pharmacol 168: 243–246

Klotz KN, Cristalli G, Grifantini M, Vittori S, Lohse MJ (1985) Photoaffinity labeling of A_1-adenosine receptors. J Biol Chem 260: 14659–14664

Libert F, Schiffmann SF, Lefort A, Parmentier M, Gérad C, Dumont JE, Vandcrhaegen JJ, Vassart G (1991) The orphan receptor cDNA RDC7 encodes an A_1 adenosine receptor. EMBO J 10: 1677–1682

Lohse MJ, Klotz KN, Lindenborn-Fotinos J, Reddington M, Schwabe U, Olsson RA (1987) 8-Cyclopentyl-1,3-dipropylxanthine (DPCPX) — a selective high affinity antagonist radioligand for A_1 adenosine receptors. Naunyn-Schmiedebergs Arch Pharmacol 336: 204–210

Lohse MJ, Klotz KN, Schwabe U, Cristalli G, Vittori S, Grifantini M (1988) 2-Chloro-N^6-cyclopentyladenosine: a highly selective agonist at A_1 adenosine receptors. Naunyn-Schmiedebergs Arch Pharmacol 337: 687–689

Londos C, Cooper DMF, Wolff J (1980) Subclasses of external adenosine receptors. Proc Natl Acad Sci USA 77: 2551–2554

Lorenzen A, Grün S, Vogt H, Schwabe U (1991) Identification of a novel high affinity adenosine binding protein from bovine striatum (submitted)

Maenhaut C, Van Sande J, Libert F, Abramowicz M, Parmentier M, Vanderhagen JJ, Dumont JE, Vassart G, Schiffmann S (1990) RDC8 codes for an adenosine A_2 receptor with physiological constitutive activity. Biochem Biophys Res Commun 173: 1169–1178

Munshi R, Linden J (1989) Co-purification of A_1 adenosine receptors and guanine nucleotide-binding proteins from bovine brain. J Biol Chem 264: 14853–14859

Nakata H (1989) Purification of A_1 adenosine receptor from rat brain membranes. J Biol Chem 264: 16545–16551

Weber R, Jones CR, Palacios JM, Lohse MJ (1990) High resolution autoradiography of A_1 adenosine receptors with [^{3}H]8-cyclopentyl-1,3-dipropylxanthine ([^{3}H]DPCPX). J Neurochem 54: 1344–1353

Williams M (1989) Adenosine: the prototypic neuromodulator. Neurochem Int 14: 249–264

Authors' address: Prof. U. Schwabe, Department of Pharmacology, University of Heidelberg, Im Neuenheimer Feld 366, D-W-6900 Heidelberg, Federal Republic of Germany

J Neural Transm (1991) [Suppl] 34: 157–162
© by Springer-Verlag 1991

Mediation by adenosine of the trophic effects exerted by the sympathetic innervation of blood vessels

W. Osswald

Department of Pharmacology and Therapeutics, Faculty of Medicine, Porto University,
Porto, Portugal

Summary. Chemical or surgical sympathetic denervation of blood vessels causes marked changes of the effector cells. Since postganglionic sympathetic cotransmission by noradrenaline and adenosine 5′-triphosphate is well established, the role of these transmitters as putative trophic factors was investigated. Whereas noradrenaline was ineffective in preventing morphological changes due to denervation, both adenosine and N-ethylcarboxamido-adenosine totally prevented them. In conscious rats, the adenosine receptor antagonist dipropylsulphophenylxantine(DPSPX) caused alterations of the blood vessel wall similar to those described for denervation. These results strongly suggest that adenosine is the trophic factor of sympathetic innervation.

Introduction

There is firm evidence for a trophic influence exerted by the sympathetic innervation of blood vessels on vascular effector cells, hypertrophy and hyperplasia going hand in hand with adrenergic denervation (for review, see Azevedo and Osswald, 1986). We have described marked morphological and functional changes of the extraneuronal cells and of the extracellular matrix after surgical denervation of the canine lateral saphenous vein and of the rabbit ear artery (Branco et al., 1984). Briefly, these changes consisted in thickening of the vessel wall, increase in dimensions of cells, indentation and reduction of heterochromatin in nuclei and accentuation of the rough endoplasmic reticulum of smooth muscle cells. Fibroblasts showed similar alterations and increased in number; the extracellular matrix became more abundant. The denervated vessel wall metabolized noradrenaline and isoprenaline in a much less efficient way than the normal one; the corticosteroid-sensitive uptake and metabolizing system became non-apparent (Branco et al., 1984).

These results were confirmed by Sarmento et al. (1987) for arteries and the heart of dogs chemically denervated with 6-hydroxydopamine (6-OHDA) and by Dimitriadou et al. (1988) for the surgically denervated rabbit cerebral artery.

Thus, the conclusion was reached that sympathetic denervation causes hypertrophic — hyperplastic changes in the vasculature and that adrenergic varicosities modulate the phenotype of the effector cell (Osswald, 1990).

Since it is now well established that cotransmission by noradrenaline and adenosine 5'-triphosphate (ATP) is the rule rather than the exception and that this applies to vascular among other tissues (as reviewed by Burnstock, 1986; Campbell, 1987; Starke et al., 1991), it was of interest to investigate the role played by the noradrenergic and the purinergic component of sympathetic innervation in the trophic action the latter exerts. The evidence for the role of the purinergic component, as based on our experiments, is summarized here.

Material and methods

Surgical sympathectomy of the lateral saphenous vein of pentobarbitone-anaesthetized dogs was effected as described earlier (Branco et al., 1984). Chemical denervation was obtained by i.v. injection of 6-OHDA ($10 \, mg \cdot kg^{-1}$ on day 0 and day 1). Constant i.v. infusions of drugs were effected with Alzet osmotic minipumps (Albino-Teixeira et al., 1990a). The noradrenaline content of denervated and control veins was determined by high pressure liquid chromatography with electrochemical detection (Albino-Teixeira et al., 1990a). The methods used for the morphological study with light and ultrastructural microscopy were those described by Azevedo et al. (1981) and Sarmento et al. (1987).

Wistar rats of 250–300 g were anaesthetized with i.p. injected pentobarbitone sodium and an Alzet minipump loaded with a 1,3-dipropyl-8-(p-sulphophenyl)xanthine (DPSPX) solution ($30 \, \mu g \cdot kg^{-1} h^{-1}$) introduced into the peritoneal cavity; for details, see Matias et al. (1991).

The saphenous veins of dogs were removed 5 days after surgical denervation or the first injection of 6-OHDA; rats were sacrificed at the end of the DPSPX infusion, i.e. 7 days after implantation of the osmotic minipump. DPSPX was obtained from RBI, Nattick, MA, USA; all other drugs used were purchased from Sigma Chemical Company (St. Louis, MO, USA). The results are expressed as arithmetic means ± standard deviation. Student's t-test was applied to differences between means (unpaired experiments) and the difference considered significant when $P < 0.05$.

Results

The effects of denervation

After surgically or chemically induced denervation, the noradrenaline content of the lateral saphenous vein of the dog was reduced to less than 10% of that of controls (namely from $2.47 \pm 0.24 \, \mu g \cdot g^{-1}$ in controls to 0.15 to $0.23 \, \mu g \cdot g^{-1}$ in denervated veins; n = 4 to 5 in each group). Morphologically, the consequences of surgical and chemical denervation (by 6-OHDA) were undistinguishable: densely degenerated nerve profiles, smooth muscle cells and fibroblasts with exuberant nuclei, rich in euchromatin, nucleoli and indentations and voluminous rough endoplasmic reticuli were regularly found in the denervated venous tissue. Moreover,

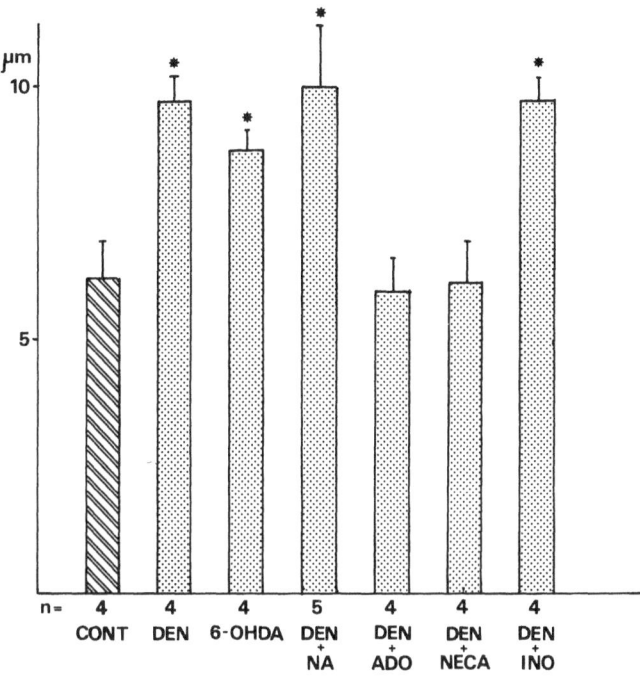

Fig. 1. Mean diameter of saphenous vein smooth muscle cells in µm (500 cells in each series; n = number of dogs; the asterisk indicates significant difference vs. controls — p < 0.001). *CONT* controls; *DEN* surgically denervated veins; *6-OHDA* chemically denervated veins; *DEN + NA* denervated veins infused with noradrenaline $0.1\,\mu g . kg^{-1}h^{-1}$; *DEN + ADO* denervated veins infused with adenosine $10\,\mu g . kg^{-1}h^{-1}$; *DEN + NECA* denervated veins infused with NECA $0.1\,\mu g . kg^{-1}h^{-1}$; *DEN + INO* denervated veins infused with inosine $10\,\mu g . kg^{-1}h^{-1}$

mast cells (which were conspicuously absent from control veins) were present in denervated veins.

Light microscopy morphometry showed that smooth muscle cell dimensions were markedly increased in the denervated veins, irrespectively of the method (surgical or chemical) used to cause denervation (Fig. 1). The size of nuclei of fibroblasts (used instead of cell dimensions due to the irregular shape of these cells) also showed a significant increase in denervated veins (from $3.76 \pm 0.90\,\mu m$ to 5.70 ± 0.52 and $5.30 \pm 0.48\,\mu m$ after surgical or chemical denervation, respectively; n = 400 nuclei for each group, P < 0.01).

The preventive effects of drugs on denervation-induced changes

Noradrenaline $(0.1\,\mu g . kg^{-1}h^{-1})$, adenosine $(10\,\mu g . kg^{-1}h^{-1})$, inosine $(10\,\mu g . kg^{-1}h^{-1})$ and 5'-N-ethylcarboxamidoadenosine (NECA; $0.1\,\mu g . kg^{-1}h^{-1})$ were infused constantly by the intravenous route for 5 days, the infusion beginning immediately after surgical denervation or the first injection of 6-OHDA. None of these drugs interfered with the denervation proper, as judged from the fact that noradrenaline content and microscopical aspect of nerve varicosities were not different from those found in denervated, not infused dogs. Noradrenaline and inosine infusions did not prevent the above described changes caused at the extraneuronal level by denervation. In sharp contrast, both adenosine and NECA infusions totally prevented these changes: smooth muscle cells and fibrobrasts showed no signs of hypertrophy or exuberant nuclear activity, mast cells were absent in the veins of animals so treated. Morphometric data confirmed this morphological appearance: in adenosine or NECA-treated animals dimensions of smooth muscle cells (and of nuclei of fibroblasts) were identical to those of control, i.e. non-denervated veins (Fig. 1).

The effects of the adenosine receptor antagonist DPSPX

Rats constantly infused with DPSPX during 7 days exhibited a number of changes ranging from increased body weight to hypertension. In the present context, only morphological changes detected at the vasculature will be reported. Morphological and morphometric studies revealed narrowing or occlusion of the lumen of small mesenterial arteries, as well as thickening and hypertrophy of smooth muscle cells in renal arteries and tail artery. The dimensions of nuclei of myocardial cells were significantly increased in comparison with saline infused controls.

Discussion

The results presented clearly show that denervation, irrespective of the method used to cause it (surgery or administration of the neurotoxin 6-OHDA), results in marked changes of the innervated cells of the saphenous vein, namely fibroblasts and smooth muscle cells, as well as in the appearance of a cell type which is absent from control veins, namely mast cells (Branco et al., 1984; Albino-Teixeira et al., 1990a). These data are in good agreement with those obtained in other tissues and/or species, as reported by Campbell et al. (1977), Fronek (1983), Sarmento et al. (1987), Dimitriadou et al. (1988) and Albino-Teixeira et al. (1989). Thus, it appears justified to state in general terms that the sympathetic nervous system exerts a modulating effect on the effector cell phenotype and that lack of this trophic action results in increased synthetic activity and hypertrophy of these cells.

In the search for the transmitter(s) responsible for this trophic effect we studied the effects of noradrenaline (which is the main functional sympathetic transmitter in the dog saphenous vein, as first demonstrated by Branodã and Guimarães, 1974) and of adenosine, the extracellularly formed metabolite of ATP, a cotransmitter of noradrenaline in a large number of tissues, including the dog saphenous vein (Flavahan and Vanhoutte, 1986). Our results strongly suggest that adenosine is the trophic factor of sympathetic innervation since it prevented completely the structural changes caused by denervation and its action was mimicked by NECA, a more stable agonist of adenosine receptors. Inosine, a metabolite of adenosine, was devoid of protective action; since NECA acts extracellularly, not being subject to cellular uptake, the site of action of both adenosine and NECA appears to be located on membrane receptors (most probably of the A2 type; neither drug is selective for these receptors, but this is the type commonly ascribed to effector cells).

This role of adenosine does not appear to be restricted to vascular tissue, since a similar trophic effect has been described on liver fibroblasts (Albino-Teixeira et al., 1990b). The results obtained in rats infused with the water soluble and non-selective antagonist DPSPX are in excellent agreement with these views. In fact, DPSPX (the action of which is due to membrane receptor blockade) caused marked changes of vascular structure, very similar to those induced by denervation, as reported above. Thus, it is tempting to speculate that the changes induced by denervation, like those caused by DPSPX, are due to lack of activation of adenosine receptors situated in vascular effector cells.

In conclusion, our results suggest an involvement of purines in the trophic effects of sympathetic innervation. Furthermore, they may ascribe a functional role to adenosine resulting from the breakdown of ATP released from vesicles, namely that of modulating the phenotype of the effector cells.

Acknowledgements

The author is indebted to his colleagues I. Azevedo, D. Branco, A. Matias, A. Albino-Teixeira and J. Polónia for permission to quote unpublished results. This study was supported by Instituto Nacional de Investigação Cientifica (FmPl).

References

Albino-Teixeirà A, Matias A, Azevedo I (1989) Prevention by adenosine of the portal tract fibroblast proliferation induced by chemical denervation. In: Ribeiro A (ed) Adenosine receptors in the nervous system. Taylor and Francis, London, pp 212

Albino-Teixeira A, Azevedo I, Branco D, Osswald W (1990a) Purine agonists prevent trophic changes caused by sympathetic denervation. Eur J Pharmacol 179: 141–149

Albino-Teixeira A, Matias A, Soares-da-Silva P, Sarmento A, Azevedo I (1990b) Effects of sympathetic denervation on liver fibroblasts: prevention by adenosine. J Auton Pharmacol 10: 181–189

Azevedo I, Osswald W (1986) Trophic role of the sympathetic innervation. J Pharmacol (Paris) 17 [Suppl 11]: 30–43

Azevedo I, Castro-Tavares J, Garrett J (1981) Ultrastructural changes in blood vessels of perinephritic hypertensive dogs. Blood Vessels 18: 110–119

Branco D, Teixeira AA, Azevedo I, Osswald W (1984) Structural and functional alterations caused at the extraneuronal level by sympathetic denervation of blood vessels. Naunyn-Schmiedebergs Arch Pharmacol 326: 302–312

Brandão F, Guimarães S (1974) Inactivation of endogenous noradrenaline released by electrical stimulation in vitro of dog saphenous vein. Blood Vessels 11: 45–54

Burnstock G (1986) The changing face of autonomic neurotransmission. Acta Physiol Scand 126: 67–91

Campbell G (1987) Cotransmission. Ann Rev Pharmacol Toxicol 27: 51–70

Campbell GR, Gibbins I, Allan I, Gannon B (1977) Effects of long term denervation on smooth muscle of the chicken expansor secundariorum. Cell Tissue Res 176: 143–156

Dimitriadou V, Aubineau P, Taxi J, Seylaz J (1988) Ultrastructural changes in the cerebral artery wall induced by long-term sympathetic denervation. Blood Vessels 25: 122–143

Fronek K (1983) Trophic effect of the sympathetic nervous system on vascular smooth muscle. Ann Biomed Eng 11: 607–615

Flavahan NA, Vanhoutte PM (1986) Sympathetic purinergic vasoconstriction and thermosensitivity in a canine cutaneous vein. J Pharmacol Exp Ther 239: 784–789

Matias A, Albino-Teixeira A, Polonia J, Azevedo I (1991) Long-term administration of 1,3-dipropyl-8-sulphophenylxanthine causes arterial hypertension. Eur J Pharmacol 193: 101–104

Osswald W (1990) Vascular hypertrophy caused by sympathetic denervation. Eur J Pharmacol 183(1): 89

Sarmento A, Soares-da-Silva P, Albino-Teixeira A, Azevedo I (1987) Effects of denervation induced by 6-hydroxydopamine on cell nucleus activity of arterial and cardiac cells of the dog. J Auton Pharmacol 7: 119–126

Starke K, von Kügelgen I, Bulloch JM, Illes P (1991) Nucleotides as cotransmitters in vascular sympathetic neuroeffector transmission. Blood Vessels 28: 19–26

Author's address: Dr. W. Osswald, Department of Pharmacology and Therapeutics, Faculty of Medicine, Porto University, 4200 Porto, Portugal

J Neural Transm (1991) [Suppl] 34: 163–169

Alpha₁- and alpha₂-adrenoceptors at different levels of the canine saphenous vein

S. Guimarães[1], **D. Moura**[1], **J. P. Nunes**[1], **M. J. Vaz-da-Silva**[1], and **J. T. Guimarães**[2]

[1]Department of Pharmacology and [2]Department of Biochemistry, Faculty of Medicine, Porto, Portugal

Summary. Presynaptic alpha₂- and postsynaptic alpha₁-adrenoceptors were compared at the distal and proximal parts of the dog saphenous vein. The results obtained show that: (1) yohimbine is more effective against postsyaptic responses to phenylephrine distally than proximally. On the contrary, WB-4101 is more effective proximally; (2) phenylephrine increases inositol monophosphate production at both levels, but the increase is more pronounced distally; (3) UK-14,304 and adrenaline reduce and yohimbine and phentolamine increase the release of ³H-noradrenaline caused by electrical stimulation at both levels. However, while adrenaline as well as the antagonists are equipotent at the two levels, UK-14,304 is more potent distally than proximally.

In conclusion, we suggest that: more alpha₁ₐ-adrenoceptors exist distally than proximally; imidazoline sites can exist at the distal level which contribute to the higher potency of UK-14,304 distally.

Introduction

The existence of two different subtypes of alpha-adrenoceptors in blood vessels was postulated on the basis of results obtained in experiments carried out in vivo (Drew and Whiting, 1979; Docherty et al., 1979). Soon afterwards, the presence of both subtypes of alpha-adrenoceptors in vessels was confirmed in in vitro experiments (De Mey and Vanhoutte, 1981; Shepperson and Langer, 1981). In 1985, Bevan and co-workers proposed that alpha-adrenoceptor responsiveness decreases with successive branching of arteries. However, Thom et al. (1985) reported that there is a reduction in alpha₂-adrenoceptor responsiveness from distal to proximal arteries, such that in larger arteries there are no alpha₂-adrenoceptor-mediated responses. These results were obtained in human limb arteries, the tissue where Flavahan et al. (1987) obtained similar results which were ascribed to an increase in the density of alpha₂-adrenoceptors at the distal level. Hence, it appears that large arteries do not possess alpha₂-adrenoceptors (Polónia et al., 1985) and that alpha₂-adrenoceptors increase in number from proximal to distal arteries (McGrath et al., 1989).

In contrast to the relative scarcity of alpha$_2$-adrenoceptors in arteries, alpha$_2$-adrenoceptors abound in most veins (De Mey and Vanhoutte, 1981; Shepperson and Langer, 1981; Constantine et al., 1982; Shoji et al., 1983; Guimarães et al., 1983, 1987). Recently, it was shown that the effectiveness of alpha$_2$-adrenoceptor activation increases from the distal to the proximal part of the veins of canine limbs, suggesting for alpha$_2$-adrenoceptors an inverse distribution to that observed in the arteries.

Furthermore, relying on differences in the effectiveness of several alpha-adrenoceptor antagonists in inhibiting the pressor responses to sympathetic stimulation and administered noradrenaline in pithed rats, Yamaguchi and Kopin (1980) suggested that the responses to sympathetic stimulation are the result of activation of alpha$_1$-adrenoceptors in the vascular neuro-effector junction, while the effects of exogenous catecholamines are mediated predominantly by alpha$_2$-adrenoceptors. Confirmatory evidence was published by Langer and Shepperson (1982) and Wilffert et al. (1982). However, at the same time, it was observed in the canine saphenous vein in vitro that noradrenaline released by electrical stimulation or by tyramine preferentially activates alpha$_2$-adrenoceptors (Guimarães et al., 1983). This was confirmed in the same tissue by Flavahan et al. (1984) and in the human saphenous vein by Docherty and Hyland (1985). Thence, there are important differences not only between arteries and veins but also between segments of each of these components of the vascular tree.

In the present study an analysis of presynaptic alpha$_2$- and postsynaptic alpha$_1$-adrenoceptors was performed at different levels of the canine saphenous vein.

Material and methods

Postsynaptic alpha$_1$-adrenoceptor-mediated responses

Segments of canine saphenous vein of about 25 mm in length were obtained from the distal and the proximal part of the vessel (Guimarães and Nunes, 1990). From each segment, a pair of helically cut strips of about 2.5×25 mm was prepared (Guimarães and Osswald, 1969). Some strips were mounted between a stationary tissue holder and a transducer and isometric contractions were recorded in grams of tension on a Harvard Universal Oscillograph. Concentration-response curves to phenylephrine were obtained by increasing the concentration of the agonist cumulatively by half-log increments in the presence of cocaine ($12 \mu mol . l^{-1}$) to inhibit uptake$_1$, hydrocortisone ($40 \mu mol . l^{-1}$) to inhibit uptake$_2$ and propranolol ($1 \mu mol . l^{-1}$) to block beta-adrenoceptors. When competitive antagonists were used, the preparations were incubated for 30 min with these drugs, which remained in the bath solution during the exposure of the tissue to the agonist.

Presynaptic alpha$_2$-adrenoceptor-mediated responses

Other strips were preincubated in the presence of 1 mmol . l^{-1} of pargyline for 30 min, exposed to $0.23 \mu mol . l^{-1}$ ^{3}H-(-)-noradrenaline during 1 h, mounted in a perifusion chamber and washed out as described by Guimarães et al. (1978). From the 110th min of washout onwards, the perifusion fluid was collected continuously in samples of 5 min. Five periods of electrical stimulation (1 Hz, 2 ms, 100 V, 5 min) were performed at 120, 165, 210, 255 and 300 min. Increasing concentrations of agonists or antagonists were added to the perifusion fluid 20 min before the 3rd, 4th and 5th periods of electrical stimulation. The overflow induced by the 2nd stimulation was taken as the control and the effect of agonists or antagonists was determined by the percentage reduction or enhancement of tritium overflow. COMT and uptake₂ were inhibited by U-0521 (50 μmol . l⁻¹) and hydrocortisone (40 μmol . l⁻¹), respectively. Uptake₁ was inhibited by cocaine (12 μmol . l⁻¹) from the 90th min on.

Inositol phosphate production

To incorporate ^{3}H-inositol into membrane phospholipids, the tissues were incubated during 3 h with this compound. Thereafter, the tissues were placed in Krebs solution containing 10 mmol . l⁻¹ lithium chloride. Then, the preparations were stimulated with phenylephrine (1 and 100 μmol . l⁻¹) for 30 min. To stop the reaction, 1.0 ml of 10‰ trichloroacetic acid was added. Subsequently, the tissues were homogenized. The supernatant of the centrifugate was neutralized with NaOH and a 1 ml sample was placed on a resin (Dowex 1-x, 400 mesh, formate form). Different ^{3}H-inositol phosphates were eluted from the column as described by Berridge et al. (1983). The fraction containing ^{3}H-inositol monophosphate (^{3}H-IP) was counted in the liquid scintillation counter and the radioactivity in it was expressed as cpm per mg of tissue. The results are expressed as the percentage increase in ^{3}H-IP by phenylephrine over that obtained in untreated controls.

Results

Postsynaptic responses

To keep their selectivity of action, both phenylephrine and the antagonists were used in concentrations as low as possible. Accordingly, the antagonist action is expressed as the shift to the right (log units) caused by the antagonist on the concentration-response curves at the EC_{25} level. As shown in Table 1, yohimbine was much more effective against the responses to phenylephrine at the distal than at the proximal level, while WB-4101 [2(2'6'-dimethoxy)-phenoxyethylamino) methylbenzodioxan] was more potent at the proximal than at the distal level.

Table 1. Shifts to the right (log units) caused by yohimbine and WB-4101 on the concentration-response curves to phenylephrine. The shifts were determined at the EC_{25} level. The values represent arithmetic means ± S.E.

Antagonist	Agonist	Proximal	Distal	n
Yohimbine (100 nmol.l⁻¹)	Phenylephrine	0.31 ± 0.02	1.32 ± 0.01*	6
WB-4101 (10 nmol.l⁻¹)	Phenylephrine	0.95 ± 0.09	0.61 ± 0.08*	5

* Significantly different from the respective proximal value (P < 0.05)

Inositol phosphate production

The results on this respect are very preliminary. Phenylephrine (1 and $100\,\mu mol\,.\,l^{-1}$) caused concentration-dependent increases of ^3H-PI production at both levels. However, the increase caused by the highest concentration of phenylephrine was 390 and 102% at the distal and the proximal level, respectively ($P < 0.05$).

Influence on the release of ^3H-(−)-noradrenaline

Either the selective alpha$_2$-adrenoceptor agonist UK-14,304 [5-bromo-6-imidazoline)-2-ylamino)-quinoxaline] or adrenaline reduced in a concentration-dependent manner the release by electrical stimulation of ^3H-(−)-noradrenaline both at proximal and distal levels (Table 2). However, while adrenaline was equipotent at both levels, UK-14,304 was about 3 times more potent distally than proximally.

Phentolamine and yohimbine increased in a concentration-dependent manner the release of ^3H-(−)-noradrenaline and each of them was equipotent at both levels (Table 3).

Table 2. Concentrations of agonists required for half-maximal inhibition (EC50) of the release of ^3H-noradrenaline elicited by electrical stimulation of the proximal and distal parts of the canine saphenous vein. Shown are geometric means and 95% confidence limits

	Proximal EC50 (nM)	Distal EC50 (nM)	n
UK-14,304	181.0 (87.1; 380.2)	61.7* (46.8; 81.3)	8
Adrenaline	98.1 (58.9; 169.8)	66.1 (21.4; 199.5)	9

*Significantly different from the proximal value ($P < 0.05$)

Table 3. Concentrations of antagonists required for a 50% increase in the release of ^3H-noradrenaline (EC50) elicited by electrical stimulation of the proximal and distal parts of the canine saphenous vein. Shown are geometric means and 95% confidence limits

	Proximal EC50 (nM)	Distal EC50 (nM)	n
Phentolamine	56.2 (31.6; 100.0)	51.3 (25.7; 79.4)	5
Yohimbine	16.6 (1.2; 229.0)	34.7 (14.1; 79.4)	3

Discussion

Postsynaptic effects

The results of the present study show that yohimbine is much more effective against the responses to phenylephrine at the distal than at the proximal level. Yohimbine clearly shows at distal level a higher antagonist affinity than that usually associated with this antagonist at classical alpha$_1$-adrenoceptors in other tissues. In 1982, Ruffolo et al. showed that alpha-adrenoceptors in the rat aorta possess properties of both alpha$_1$- and alpha$_2$-adrenoceptors, since both prazosin and yohimbine are very potent against phenylephrine in this tissue. The alpha-adrenoceptors mediating responses to phenylephrine in the distal portion of the canine saphenous vein appear to be like those described by these authors.

Alpha$_1$-adrenoceptors have been recently subclassified from functional studies (Han et al., 1987) into alpha$_{1A}$ and alpha$_{1B}$. The high antagonist potency of WB-4101 on contractions to phenylephrine agrees with the reported affinity of WB-4101 at these sites as defined by binding studies (Minneman et al., 1988; Michel et al., 1989) and is consistent with the presence of alpha$_{1A}$-sites at both levels of the canine saphenous vien (Hicks et al., 1991). However, as our results show, WB-4101 is more potent against responses to phenylephrine proximally than distally, apparently indicating that more alpha$_{1A}$-adrenoceptors exist proximally than distally.

Additionally, the present results show that phenylephrine causes an increase in ^{3}H-IP production, indicating that inositol phosphates are involved in the responses to this agonist both in the proximal and distal portions of the vein. However, the increase in ^{3}H-IP formation elicited by phenylephrine is more pronounced distally than proximally, indicating, according to current theory (Han et al., 1987; Michel et al., 1990), that more alpha$_{1B}$-adrenoceptors exist distally than proximally. Thence, there is a good agreement between functional and biochemical approaches supporting the view that more alpha$_{1B}$-adrenoceptors exist distally than proximally, the opposite occurring for alpha$_{1A}$-adrenoceptors.

Presynaptic effects

UK-14,304 — an imidazoline — is significantly more potent at inhibiting noradrenaline release at the distal than at the proximal level of the canine saphenous vein. Since the noradrenaline content of the proximal part of the vein is higher than that of the distal one (Pereira et al., 1991), this differential effect might be ascribed to a higher synaptic concentration of the transmitter during electrical stimulation at the proximal than at the distal level (Starke, 1972; Gonçalves et al., 1989). However, adrenaline — a phenylethylamine — caused identical reductions in noradrenaline release at both levels, and phentolamine and yohimbine caused similar increases in noradrenaline overflow in both parts of the vessel. Although preliminary, these results suggest the possibility that some imidazoline sites exist at the distal level which contribute to the higher potency of UK-14,304 at this level.

Acknowledgements

This work was supported in part by JNICT (Junta Nacional de Investigação Científica e Tecnológica) — Projecto n° PMCT/SAU/236/90 and in part by INIC (Instituto Nacional de Investigação Científica) — FmP1.

References

Bevan JA, Bevan RD, Laher I (1985) Role of alpha-adrenoceptors in vascular control. Clin Sci 68 [Suppl 10]: 83s–89s

Berridge MJ, Dawson RMC, Downes CP, Heslop, JP, Irvine RF (1983) Changes in the levels of inositol phosphates after agonist-dependent hydrolysis of membrane phosphoinositides. Biochem J 212: 473–482

Constantine JW, Lebel W, Archer R (1982) Functional postsynaptic α_2- but not α_1-adrenoceptors in dog saphenous vein exposed to phenoxybenzamine. Eur J Pharmacol 85: 325–329

De Mey JC, Vanhoutte PM (1981) Uneven distribution of postjunctional alpha$_1$- and alpha$_2$-like adrenoceptors in canine arterial and venous smooth muscle. Circ Res 48: 875–884

Docherty JR, Hyland L (1985) Evidence for neuro-effector transmission through postjunctional α_2-adrenoceptors in human saphenous vein. Br J Pharmacol 84: 573–576

Docherty JR, MacDonald A, McGrath JC (1979) Further subclassification of α-adrenoceptors in the cardiovascular system, vas deferens and anococcygeus of the rat. Br J Pharmacol 67: 421P–422P

Drew GM, Whiting SB (1979) Evidence for two distinct types of postsynaptic α-adrenoceptors in vascular smooth muscle in vivo. Br J Pharmacol 67: 207–215

Flavahan NA, Rimele JP, Cooke JP, Vanhoutte PM (1984) Characterization of postjunctional alpha$_1$- and alpha$_2$-adrenoceptors activated by exogenous or nerve released norepinephrine in the canine saphenous vein. J Pharmacol Exp Ther 230: 399–705

Flavahan NA, Cooke JP, Shepherd JT, Vanhoutte PM (1987) Human postjunctional alpha$_1$- and alpha$_2$-adrenoceptors: differential distribution in arteries of the limbs. J Pharmacol Exp Ther 241: 361–365

Gonçalves J, Carvalho F, Guimarães S (1989) Uptake inhibitors do not change the effect of imidazoline α_2-adrenoceptor agonists on transmitter release evoked by single pulse stimulation in mouse vas deferens. Naunyn-Schmiedebergs Arch Pharmacol 339: 288–292

Guimarães S, Osswald W (1969) Adrenergic receptors in the veins of the dog. Eur J Pharmacol 5: 133–140

Guimarães S, Nunes JP (1990) The effectiveness of α_2-adrenoceptor activation increases from the distal to the proximal part of the veins of canine limbs. Br J Pharmacol 101: 387–393

Guimarães S, Brandão F, Paiva MQ (1978) A study of the adrenoceptor-mediated feedback by using adrenaline as a false transmitter. Naunyn-Schmiedebergs Arch Pharmacol 305: 185–188

Guimarães S, Paiva MQ, Polónia JJ (1983) Alpha$_1$- and alpha$_2$-adrenoceptors of the dog saphenous vein and their relation to the sympathetic nerve terminals. Prog Neuropsychopharmacol (Abstr) [Suppl]: 155

Guimarães S, Paiva MQ, Moura D (1987) Alpha$_2$-adrenoceptor-mediated responses to so-called selective alpha$_1$-adrenoceptor agonists after partial blockade of alpha$_1$-adrenoceptors. Naunyn-Schmiedebergs Arch Pharmacol 335: 397–402

Han L, Abel PW. Minneman KP (1987) Alpha$_1$-adrenoceptor subtypes linked to different mechanisms for increasing intracellular Ca^{2+} in smooth muscle. Nature 329: 333–335

Hicks PE, Barras M, Herman G, Mauduit P, Armstrong JM, Rossignol B (1991) α-Adrenoceptor subtypes in dog saphenous vein that mediate contraction and inositol phosphate production. Br J Pharmacol 102: 151–161

Langer SZ, Shepperson NB (1982) Recent developments in vascular smooth muscle pharmacology: the post-synaptic α$_2$-adrenoceptor. Trends Pharmacol Sci 3: 440–444

McGrath JC, Brown CM, Wilson VG (1989) Alpha-adrenoceptors: a critical review. Med Res Rev 9: 407–533

Michel MC, Hanft G, Gross G (1990) Alpha$_{1B}$-but not alpha$_{1A}$-adrenoceptors mediate inositol phosphate generation. Naunyn-Schmiedebergs Arch Pharmacol 341: 385–387

Minneman LP, Han C, Abel PW (1988) Comparison of alpha$_1$-adrenoceptor subtypes distinguished by chloroethylclonidine and WB4101. Mol Pharmacol 33: 509–514

Pereira O, Moura D, Nunes P, Vaz-da-Silva MJ, Guimarães S (1991) Involvement of α$_1$- and α$_2$-adrenoceptors in the responses of proximal and distal segments of the canine saphenous vein to exogenous and endogenous noradrenaline. Naunyn-Schmiedebergs Arch Pharmacol 343: 616–622

Polónia JJ, Paiva MQ, Guimarães S (1985) Pharmacological characterization of postsynaptic α-adrenoceptor subtypes in five different dog arteries in vitro. J Pharm Pharmacol 37: 205–208

Ruffolo RR, Jr. Waddel JE, Yaden EC (1982) Heterogeneity of postsynaptic alpha adrenergic receptors in mammalian aorta. J Pharmacol Exp Ther 221: 309–314

Shepperson NB, Langer SZ (1981) The effects of the 2-aminotetrahydronaphtalene derivative M7 a selective α$_2$-adrenoceptor agonist in vitro. Naunyn-Schmiedebergs Arch Pharmacol 318: 10–13

Shoji T, Tsuru H, Shigei T (1983) A regional difference in the distribution of postsynaptic alpha-adrenoceptor subtypes in canine veins. Naunyn-Schmiedebergs Arch Pharmacol 324: 246–255

Starke K (1972) Alpha sympathomimetic inhibition of adrenergic and cholinergic transmission in the rabbit heart. Naunyn-Schmiedebergs Arch Pharmacol 274: 18–45

Thom S, Calvete J, Hayes R, Martin G, Sever P (1985) Human vascular smooth muscle responses mediated by alpha$_2$ mechanisms in vivo and in vitro. Clin Sci 68: 147s–150s

Wilffert B, Timmermans PBMWM, Van Zwieten PA (1982) Extrasynaptic location of alpha-2 and noninnervated beta-2 adrenoceptors in the vascular system of the pithed normotensive rat. J Pharmacol Exp Ther 221: 762–768

Yamaguchi I, Kopin IJ (1988) Differential inhibition of alpha-1 and alpha-2 adrenoceptor-mediated pressor responses in pithed rats. J Pharmacol Exp Ther 214: 275–287

Authors' address: Dr. S. Guimarães, Department of Pharmacology, Faculty of Medicine, 4200 Porto, Portugal

J Neural Transm (1991) [Suppl] 34: 171–177

Pre- and postsynaptic alpha-2 adrenoceptors as target for drug discovery

S. Z. Langer and **I. Angel**

Department of Biology, Synthélabo Recherche (LERS), Paris, France

Summary. The presynaptic terminal autoreceptors which modulate the release of noradrenaline through a negative feed-back mechanism correspond to the α_2-subtype of adrenoceptors. These receptors are also present postsynaptically in the pancreatic islets were they mediate inhibition of the glucose-induced release of insulin. The sympathetic innervation of the pancreatic islets involves α_2-adrenoceptors both presynaptically as well as postsynaptically. SL 84.0418 is a novel α_2-adrenoceptor antagonist with preferential effects in the periphery and with at least a 10-fold higher selectivity ratio between α_2 and α_1-adrenoceptors when compared with idazoxan. SL 84.0418 antagonizes the hyperglycemia and the inhibition of insulin release induced by the α_2-adrenoceptor agonist UK 14304. The administration of SL 84.0418 significantly reduces the glucose evoked hyperglycemia in several species including man. It is proposed that SL 84.0418 may represent a useful and novel hypoglycemic drug in the treatment of type II diabetes.

Introduction

The concept that neurotransmitters can regulate their own release through presynaptic terminal autoreceptors was first developed for noradrenaline and subsequently extended to other transmitters (Langer, 1974, 1981; Starke et al., 1989).

In the noradrenergic system the negative feed-back modulation of transmitter release developed in parallel with the pharmacological evidence for two subtypes of α-adrenoceptors as defined by a different profile of affinity and relative order of potencies for agonists and for antagonists (Langer, 1981).

The α_1-adrenoceptor subtype is present in vascular smooth muscle, where it mediates vasoconstrictor responses. The presynaptic adrenoceptor that mediates the inhibition of the release of noradrenaline corresponds to the α_2-adrenoceptor subtype. Postsynaptic α_2-adrenoceptors are also present in vascular smooth muscle, in adipocytes, platelets and in the insulin secreting β-cells of the pancreas. Recent molecular biology studies as well as studies with functional and receptor binding techniques revealed that subtypes of α_1 as well as α_2-adrenoceptors exist (Minneman, 1988; Bylund, 1988).

The growing interest in the development of selective α_2-adrenoceptor antagonists was reflected in the synthesis of several compounds which are more selective than yohimbine and rauwolscine as α_2-adrenoceptor antagonists. Idazoxan (Dettmar et al., 1983; Doxey et al., 1984) is a potent and selective α_2-adrenoceptor antagonist in the periphery and in the central nervous system.

In the present article we describe the pharmacological profile of SL 84.0418, a pyrrolo-indole derivative with high selectivity for α_2-adrenoceptors in the periphery (Langer et al., 1990; Angel et al., 1991a,b). This α_2-adrenoceptor antagonist is presently tested in man as a potential antidiabetic for the treatment of non-insulin dependent diabetes mellitus.

Materials and methods

The following isolated organ preparations were employed: twitch response of the isolated rat vas deferens (Hicks et al., 1985); dog saphenous vein (Rhodes and Waterfall, 1987); rabbit pulmonary artery (Schoemaker et al., 1989) and rat hypothalamic slices prelabelled with ^{3}H-noradrenaline (Galzin et al., 1984).

Insulin release from rat pancreatic islets was studied using the method described by Niddam et al. (1990), and the α_2-adrenoceptor mediated hyperglycemia in mice was carried out according to the method described by Angel et al. (1990a).

Results

The structure of SL 84.0418, a pyrrolo-indole derivative, is shown in Fig. 1. In the rat vas deferens SL 84.0418 antagonized competitively the inhibition of the twitch response induced by clonidine with a pA_2 value similar to that obtained with idazoxan (Table 1).

The pA_2 value for SL 84.0418 was similar when postsynaptic α_2-adrenoceptors were studied in the dog saphenous vein (Table 1). The potency of SL 84.0418 to antagonize α_1-adrenoceptor mediated responses of the rabbit pulmonary artery was rather low, with a pA_2 value of 5 (Table 1).

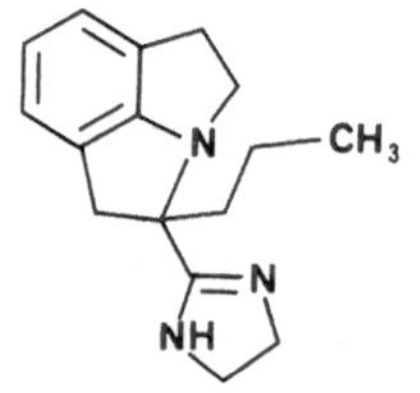

SL 84.0418

Fig. 1. Structure of SL 84.0418 (2-(4, 5-dihydro-1h-imidazol-2-yl)-1, 2, 4, 5-tetrahydro-2 propylpyrrolo [3, 2, 1-hi]-indole hydrochloride)

Table 1. Comparison of the alpha-2 and alpha-1 antagonist properties of SL 84.0418 and idazoxan

Drug	Alpha-2 antagonism		Alpha-1 antagonism	Alpha-1/ Alpha-2 selectivity ratio
	Pre-synaptic pA_2 vs clonidine	Post-synaptic pA_2 vs BHT933	Post-synaptic pA_2 vs phenylephrine	
SL 84.0418	8.28 ± 0.23	7.97 ± 0.04	4.97 ± 0.10	1,995/1,000
Idazoxan	8.34 ± 0.03	7.28 ± 0.04	6.08 ± 0.20	182/15.8

Drug antagonism at alpha-1 and alpha-2 adrenoceptor was determined as described in Methods. Shown are pA_2 values $\pm$S.E.M. of at least 3 experiments. The selectivity ratio represents the antilog of the pA_2 for pre- and postsynaptic alpha-2 antagonism minus the pA_2 for postsynaptic alpha-1 antagonist effects

Therefore the selectivity ratio of SL 84.0418 is 10-fold higher than that of idazoxan when the potencies to block α_2-adrenoceptors and α_1-adrenoceptors are compared (Table 1).

Like idazoxan, SL 84.0418 blocks presynaptic α_2-autoreceptors in rat hypothalamic slices and thus enhances the calcium-dependent, stimulation-evoked release of ^{3}H-noradrenaline without affecting the spontaneous outflow of radioactivity. The incubation with SL 84.0418 (0.1 or 1.0 µM) significantly increases the fractional ^{3}H-noradrenaline release from 1.00 ± 0.06 (n = 10) to 1.45 ± 0.02 (n = 3) and 2.02 ± 0.12 (n = 4), respectively.

As shown in Fig. 2 (top) exposure to SL 84.0418 did not on its own modify the glucose induced enhancement of insulin release from isolated rat pancreatic islets. However, SL 84.0418, concentration-dependently antagonized the inhibitory effects of the α_2-adrenoceptor agonist, UK 14304, on insulin release (Fig. 2, bottom).

In mice, the administration of the α_2-adrenoceptor agonist, UK 14304, produces a dose-dependent hyperglycemic response which is associated with an inhibition of the release of insulin (Angel et al., 1988, 1990a). Figure 3 shows that SL 84.0418 antagonized dose-dependently the hyperglycemia induced by UK 14304 in mice.

Discussion

The novel α_2-adrenoceptor antagonist SL 84.0418 is the most selective compound so far reported in this pharmacological class. The ratio of selectivity between blockade of α_2 and α_1-adrenoceptors is 10-fold higher than for idazoxan and other recently developed α_2-adrenoceptor antagonists. In addition SL 84.0418 possesses high specificity for α_2-adrenoceptors since it has very low or no affinity for a variety of receptors to different neurotransmitters, ion channels and neuronal transporters (Angel et al., 1991a).

It appears that SL 84.0418 is an antagonist at both the presynaptic and the postsynaptic α_2-adrenoceptors with similar potency. In support of this

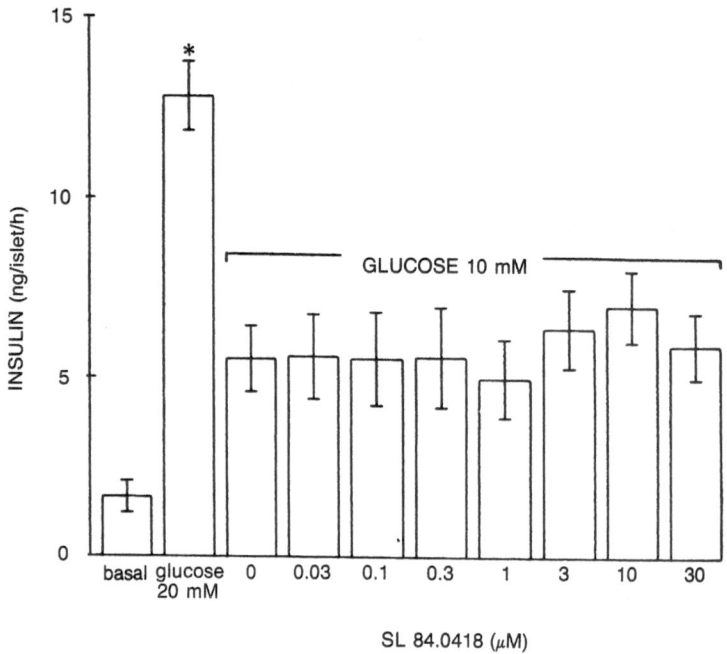

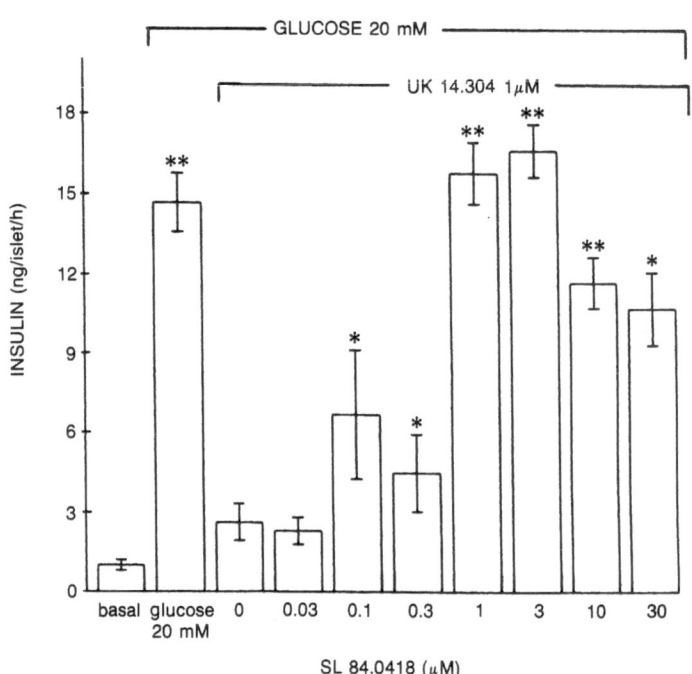

Fig. 2. Effect of SL 84.0418 on glucose induced insulin release (top) and its antagonism of the inhibition of insulin release induced by UK 14304 (bottom). Data represents means ± S. E. M. of insulin release from batches of 3 isolated islets (n = 10–15), incubated either with low (basal, 2.8 mM), intermediate (10 mM) or high (20 mM) glucose concentration and in the presence or absence of SL 84.0418 and UK 14304, as indicated. *p < 0.05, **p < 0.01, compared to the respective control, test of Dunnett

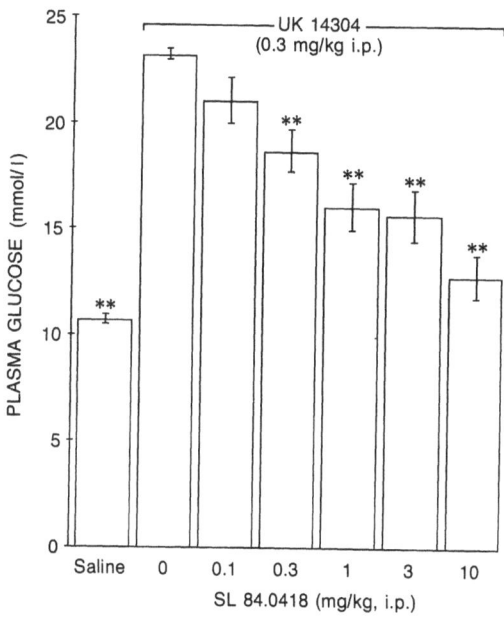

Fig. 3. Antagonism by SL 84.0418 of the hyperglycemia induced by the alpha-2 adrenoceptor agonist UK 14304. Alpha-2 adrenoceptor mediated hyperglycemia was studied in the mouse, according to Angel et al. (1990a). Mice (n = 6–28 per group) were administered the indicated dose of SL 84.0418, 20 minutes before UK 14304 (0.3 mg/kg ip). Blood was collected 30 minutes later and glucose analysed spectrophotometrically. **p < 0.01, compared to the UK 14304 group, test of Dunnett

view, it was shown that SL 84.0418 enhances the release of endogenous noradrenaline elicited by sympathetic nerve stimulation in the in situ blood perfused dog pancreas (Duval et al., 1991).

One of the interesting pharmacologcal properties of SL 84.0418 is its preferential effects on peripheral α_2-adrenoceptors with little or no activity as an antagonist of central α_2-adrenoceptors. While SL 84.0418 antagonizes the hypertensive effects of UK 14304 in the pithed rat, it fails to antagonize the hypotension and bradycardia induced by intracerebroventricularly administered clonidine (Angel et al., 1991b). Another indication for the preferential peripheral effect of SL 84.0418 is the inability of this compound to antagonize the clonidine-induced hypomotility even at doses 60-fold higher than those required to block by 50% the adrenaline induced hyperglycemia (Angel et al., 1991b). In contrast, under the same experimental conditions idazoxan is equipotent to antagonize the clonidine-induced hypomotility and the adrenaline induced hyperglycemia (Angel et al., 1991b). The preferential blockade of peripheral α_2-adrenoceptors by SL 84.0418 may be due to the fact that this compound does not penetrate readily through the blood brain barrier.

Alternatively these results may reflect the specificity of SL 84.0418 for α_2-adrenoceptor subtypes which are present in the periphery, and which may be different from those in the central nervous system.

The α_2-adrenoceptor antagonist SL 84.0418 antagonized the inhibition of insulin release obtained by UK 14304 under in vitro conditions and also the hyperglycemia induced by the α$_2$-adrenoceptor agonist under in vivo conditions. In the monkey, the oral administration of SL 84.0418 reduced in a dose-dependent manner the hyperglycemia induced by an oral glucose load (Hulbron et al., 1990). This antihyperglycemic effect of SL 84.0418 occurred at doses which did not affect basal insulin levels but increased significantly the release of insulin elicited by the glucose challenge (Hulbron et al., 1990).

Recent results indicate that similar to the monkey and other animal species, the oral administration of SL 84.0418 to man reduces the hyperglycemia induced by an oral glucose load (Morselli, personal communication). Since these anti-hyperglycemic effects in man occur at doses which are devoid of cardiovascular and other side effects (Bergougnan et al., 1990), addition studies are currently under way to establish if SL 84.0418 produces the expected hypoglycemic effects in patients with type II diabetes.

Two additional properties of SL 84.0418 which are associated with α$_2$-adrenoceptor blockade may be useful for an orally active antidiabetic. The first involves the antagonism of the adrenaline-induced platelet aggregation (Angel et al., 1991a) which may be beneficial in preventing or reducing the vascular complications of diabetes. The second property is related to the blockade of α$_2$-adrenoceptors in adipocytes which are linked to antilipolytic effects (Lafontan et al., 1985). The antagonism by SL 84.0418 of α$_2$-adrenoceptor-mediated effects in adipocytes (Angel et al., 1990b) may be beneficial in type II diabetes, which is frequently associated with obesity.

In summary, SL 84.0418 is a novel selective α$_2$-adrenoceptor antagonist, with preferential effects in the periphery. The pharmacological profile of SL 84.0418 is compatible with the proposal that this compound may be effective in the treatment of type II, non insulin-dependent diabetes.

References

Angel I, Bidet S, Langer SZ (1988) Pharmacological characterization of the hyperglycemia induced by alpha-2 adrenoceptor agonists. J Pharmacol Exp Ther 246: 1098–1103

Angel I, Niddam R, Langer SZ (1990a) Involvement of alpha-2 adrenergic receptor subtypes in hyperglycaemia. J Pharmacol Exp Ther 254: 877–882

Angel I, Schoemaker H, Azerhad R, Bidet S, Duval N, Langer SZ (1990b) SL 84.0418: a new alpha-2 antagonist with antihyperglycemic and lipolytic activity. Int J Obesity 14 [Suppl 2]: 50

Angel I, Schoemaker H, Arbilla S, Galzin AM, Berry C, Niddam R, Pimoule C, Sevrin M, Wick A, Langer SZ (1991a) SL 84.0418: a novel, potent and selective alpha-2 adrenoceptor antagonist. I. In vitro pharmacological profile. J Pharmacol Exp Ther (submitted)

Angel I, Grosset A, Perrault G, Schoemaker, H, Langer SZ (1991b) SL 84.0418: a new, potent and selective alpha-2 adrenoceptor antagonist with preferential peripheral activity. II. In vivo pharmacological profile. J Pharmacol Exp Ther (submitted)

Bergougnan L, Rosenzweig P, Duchier J, Cournot A, Berlin I, Morselli PL (1990) SL

84.0418: clinical and cardiovascular tolerance in healthy young volunteers of a new alpha-2 antagonist with anti-hyperglycaemic properties. Eur J Pharmacol 183: 1015–1016

Bylund DB (1988) Subtypes of α_2-adrenoceptors: pharmacological and molecular biological evidence converge. TIPS 9: 356–361

Dettmar PW, Lynn AG, Tulloch IF (1983) Neuropharmacological studies in rodents on the action of RX781094, a new selective α_2-adrenoceptor antagonist. Neuropharmacology 22: 729–739

Doxey JC, Roach AG, Strachan DA, Virdee NK (1984) Selectivity and potency of 2-alkyl analogues of the α_2-adrenoceptor antagonist idazoxan (RX781094) in peripheral systems. Br J Pharmacol 83: 713–722

Duval N, Angel I, Eon MT, Oblin A, Langer SZ (1991) Involvement of alpha-2 and beta-adrenoceptors in insulin release induced by pancreatic nerve stimulation. J Pharmacol Exp Ther (submitted)

Galzin AM, Moret C, Langer SZ (1984) Evidence that exogenous but not endogenous norepinephrine activates the presynaptic alpha$_2$ adrenoceptors on serotonergic nerve endings in the rat hypothalamus. J Pharmacol Exp Ther 228: 725–732

Hicks PE, Langer SZ, Macrae AD (1985) Differential blocking actions of idazoxan against the inhibitory effects of 6-fluoronoradrenaline and clonidine in the rat vas deferens. Br J Pharmacol 86: 141–150

Hulbron G, Angel I, Oblin A, Friedmann JC, Langer SZ (1990) Antihyperglycaemic effects of the new alpha-2 antagonist SL 84.0418 in the primate (macaca fascicularis) in comparison with idazoxan. Eur J Pharmacol 183: 995–996

Lafontan M, Berlan M, Carpene C (1985) Fat cell adrenoceptors: inter and intraspecific differences and hormone regulation. Int J Obesity 9 [Suppl]: 117–127

Langer SZ (1974) Presynaptic regulation of catecholamine release. Biochem Pharmacol 23: 1793–1800

Langer SZ (1981) Presynaptic regulation of the release of catecholamines. Pharmacol Rev 32: 337–362

Langer SZ, Schoemaker H, Angel I, Arbilla S, Pimoule C, Grosset A, Perrault G, Sevrin M, Wick A (1990) SL 84.0418: a new, potent and selective alpha-2 adrenoceptor antagonist with peripheral activity. Eur J Pharmacol 183(3): 802

Minneman KP (1988) α_1-Adrenergic receptor subtypes, inositol phosphates, and sources of cell Ca^{2+}. Pharmacol Rev 40(2): 87–119

Niddam R, Angel I, Bidet S, Langer SZ (1990) Pharmacological characterization of alpha-2 adrenergic receptor subtype involved in the release of insulin from isolated rat pancreatic islets. J Pharmacol Exp Ther 254: 883–887

Rhodes KF, Waterfall JF (1987) The effects of some α-adrenoceptor antagonists on the responses of the canine saphenous vein to B-HT 933, UK-14304 and methoxamine. Naunyn-Schmiedebergs Arch Pharmacol 335: 261–268

Schoemaker H, Blanchard H, Pimoule C, Lefevre-Borg F, Manoury, P, Jardin A, Langer SZ (1989) Characterization of the effects of alfuzosin on α_1-adrenoceptors in the genito-urinary tract. In: Prostate et Alpha-Bloquants. Excerpta Medica, Amsterdam, pp 328–343

Starke K, Göthert M, Kilbinger H (1989) Modulation of neurotransmitter release by presynaptic autoreceptors. Physiol Rev 69: 864–989

Authors' address: Dr. S. Z. Langer, Department of Biology, Synthélabo Recherche (LERS), 58/60, rue de la Glacière, BP 7, F-75622 Paris Cédex 13, France

J Neural Transm (1991) [Suppl] 34: 179–185

Cyclic AMP and adaptive supersensitivity in guinea pig atria

W. W. Fleming and **D. A. Taylor**

Department of Pharmacology and Toxicology, West Virginia University,
Morgantown, WV, U.S.A.

Summary. Supersensitivity was induced by injection of guinea pigs with reserpine, 0.1 mg/kg/day for 7 days. This treatment induced chronotropic supersensitivity of isolated right atria to isoproterenol but not to forskolin. The pretreatment induced supersensitivity to the cyclic AMP-generating effects of isoproterenol or forskolin in left, but not right, atria. These results are discussed in reference to the extensive literature describing supersensitivity in the guinea-pig heart, including the possible role of the adenylyl cyclase system.

Introduction

Neurones, muscle cells and exocrine gland cells undergo compensatory changes in sensitivity in response to chronic decreases in the physiological stimulus they receive. This phenomenon of adaptive supersensitivity has been extensively reviewed (Fleming et al., 1973; Fleming, 1976, 1981; Fleming and Westfall, 1988). Multiple mechanisms, including increases in cholinoceptor density, a partial depolarization and possible changes in calcium homeostasis contribute to adaptive supersensitivity in skeletal muscle. In certain smooth muscle, the evidence points to a partial membrane depolarization, secondary to changes in the Na^+, K^+ pump, as the cellular basis for supersensitivity.

In contrast, the identification of the cellular mechanism of adaptive supersensitivity in guinea pig atria has been elusive (Fleming, 1984). Neither changes in receptors (Torphy et al., 1982) nor electrophysiology (Schulz et al., 1984) appear to contribute. Nevertheless, the supersensitivity in guinea-pig atria is highly specific for beta adrenoceptor agonists (Torphy et al., 1982). In the left atrium, supersensitivity has also been demonstrated to substances, including forskolin, which act on the transduction process beyond the level of the beta adrenoceptor (Hawthorn et al., 1987).

In the present experiments, supersensitivity has been induced in right and left atria by chronic pretreatment of guinea pigs with reserpine. Two questions are addressed. (1) Is there chronotropic supersensitivity to forskolin in right atria? (2) Is the agonist-induced accumulation of cyclic AMP altered in right or left atria in association with supersensitivity.

Methods

Animals and pretreatment

Adult male guinea pigs (250–400 gms) were randomly assigned to either control (untreated) or reserpine (0.1 mg/kg/day × 7, s.c.) treated groups. This schedule produces 95% depletion of myocardial catecholamines within 24 h (Crout et al., 1962) and supersensitivity at 7 days (Torphy et al., 1982). Following pretreatment, the animals were stunned, exsanguinated and the right and left atria removed for subsequent experiments.

Chronotropic concentration-response curves

Atria were placed in water-jacketed organ baths filled with Chenoweth-Koelle (CK) solution maintained at 37°C and bubbled with 95% O_2-5% CO_2. The CK solution contained (mM): NaCl (120); KCl (5.6); $CaCl_2$ (2.2); $MgCl_2$ (2.1); $NaHCO_3$ (25) and glucose (10). Frequency of contraction was monitored from Grass FT.03 force-displacement transducers connected to a Grass polygraph. A resting tension of 1 g was applied and a period of one hour was allowed for equilibration with replacement of the CK solution at 10–15 minute intervals. The tissues were then challenged with cumulatively increasing concentrations of isoproterenol or forskolin and each response was calculated as a percentage of the maximum increase in rate produced by that agonist in that tissue. Changes in sensitivity to a given agonist were determined by comparing geometric mean EC_{50} values for atria obtained in each treatment group (Fleming et al., 1973).

Cyclic AMP determination

Right atria were removed, placed in organ baths in CK solution bubbled with 95% O_2-5% CO_2 and maintained at 37°C. Right atria were prepared for monitoring isometric tension in a manner similar to that described above. Left atria were placed in Petri dishes containing CK solution bubbled with 95% O_2-5% CO_2 and maintained at 37°C. Tissues were allowed to equilibrate for 1 h during which time the CK solution was changed at 10–15 min intervals. Each pair of atria was exposed simultaneously to a single, identical concentration of agonist. The inotropic response of the right atrium was monitored on a Grass polygraph. At the peak of the response, both atria were rapidly removed and frozen using clamps which had been cooled in liquid nitrogen and stored at −70°C.

Frozen tissues were thawed and homogenized in 3% perchloric acid. Potassium bicarbonate was added until pH was 5.5–6.0 and samples centrifuged at 2000 xg for 20 min to precipitate proteins. Following centrifugation, the supernatants were removed and an aliquot used for determination of cyclic AMP content. The pellet was resuspended in 1N NaOH and used for determination of protein content by the method of Lowry (1951). Cyclic AMP content in each tissue was determined using a radio-immunoassay (Diagnostic Products, Inc.) and liquid scintillation spectrometry. Comparisons of cyclic AMP levels among the different groups was made using analysis of variance and Student's "t" test for unpaired samples.

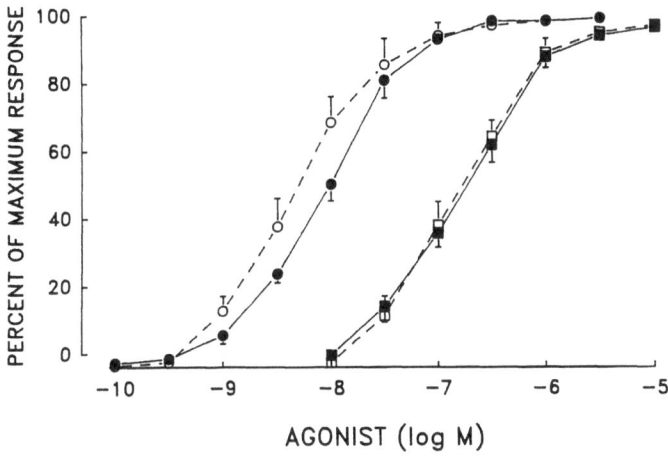

Fig. 1. Chronotropic concentration-response curves to isoproterenol (circles) and forskolin in right atria from control (solid symbols, solid lines) and reserpine-pretreated (open symbols, broken lines) guinea pigs. Vertical bars represent SEM. Numbers of preparations, per group = 4 each for isoproterenol; 12 control and 9 reserpine treated for forskolin

Results

Figure 1 demonstrates a significant ($p < 0.05$) shift to the left of the chronotropic concentration-response curve for isoproterenol in right atria from guinea pigs pretreated with reserpine. The shift (1.9-fold) is similar to the shift our laboratory has consistently obtained in this preparation from this schedule of reserpine pretreatment (Taylor et al., 1976; Torphy et al., 1982). In contrast, Fig. 1 indicates that pretreatment with reserpine had no effect on the chronotropic sensitivity to forskolin.

Isoproterenol induced concentration-dependent increases in cyclic AMP in both right and left atria (Fig. 2). In the right atria from reserpine-pretreated animals, there was a tendency for isoproterenol-induced levels of cyclic AMP to be higher. However, the differences were not statistically significant. In left atria from reserpine-pretreated guinea pigs, the cyclic AMP generated in the presence of 10^{-7} isoproterenol was more than doubled by the pretreatment, such that it was approximately a maximal response. Forskolin also induced concentration-related increases in cyclic AMP (Fig. 3). The increases were significantly greater in preparations from animals pretreated with reserpine only in the left atria. Pretreatment was associated also with a significant increase in basal concentrations of cyclic AMP in the left atria.

Discussion

The development of cardiac supersensitivity induced by depletion of catecholamines has been reported to occur in cats (Fleming and

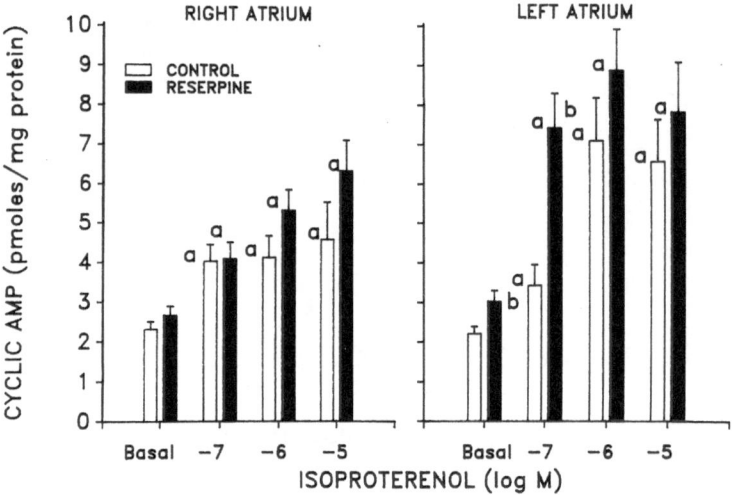

Fig. 2. Basal and isoproterenol-stimulated cyclic AMP concentrations in right and left atria from control and reserpine-pretreated guinea pigs. Vertical lines represent SEM. Each column is the mean value from at least 7 atria. **a** Indicates significant differences from basal values; **b** indicates significant difference from control (paired open column)

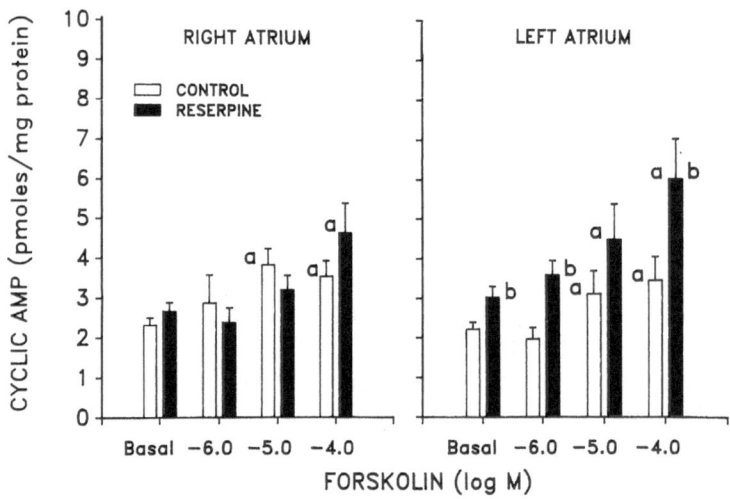

Fig. 3. Basal and forskolin-generated cyclic AMP concentrations in right and left atria from control and reserpine-pretreated guinea pigs. Vertical lines represent SEM. Each column is the mean value from at least 6 atria. **a** Indicates significant differences from basal values; **b** indicates significant difference from control (paired open column)

Trendelenburg, 1961), dogs (Trendelenburg and Gravenstein, 1958; Westfall and Fleming, 1968a), guinea pigs (Westfall and Fleming, 1968b), rabbits (Tenner and Carrier, 1978) and rats (Rice et al., 1987). Other procedures, including chronic administration of beta-adrenoceptor antagonists (Tenner, 1983) and surgical sympathetic denervation (Goto et al., 1985) also induce adaptive supersensitivity in cardiac tissues. Nevertheless, identification of the mechanism(s) of supersensitivity in myocardial cells has been difficult. Among the difficulties are apparent species differences (Fleming, 1984).

Regional differences occur in regard to supersensitivity in guinea-pig hearts. Both chronotropic and inotropic supersensitivity in right atria are highly specific for beta-adrenoceptor agonists such as norepinephrine and isoproterenol (Taylor et al., 1976; Torphy et al., 1982). Chronic treatment of guinea pigs with reserpine does not induce supersensitivity of the right atrium to calcium, histamine, aminophylline, pilocarpine (Torphy et al., 1982) or to forskolin (present results). There is no indication of any change in beta-adrenoceptor density or affinity in right atria (Torphy et al., 1982). Results presented here indicate a tendency for increases in cyclic AMP levels induced by isoproterenol to be enhanced in right atria from reserpine-pre-treated guinea pigs, but the differences were not significant. Stimulation of cyclic AMP formation by forskolin was not enhanced in right atria of pretreated animals.

In left atria of guinea pigs, pretreatment with reserpine induces inotropic supersensitivity to beta-adrenoceptor agonists, forskolin and GppNHp, a more stable analog of GTP (Hawthorn et al., 1987). The supersensitivity to forskolin and GppNHp suggests an alteration in the transduction mechanism beyond the level of the receptor. Consistent with that hypothesis are the present results showing that pretreatment with reserpine induces supersensitivity to the ability of both isoproterenol and forskolin to increase left atrial levels of cyclic AMP. Pretreatment also caused small but significant increases in basal content of cyclic AMP, in left atria.

The results in the left atrium could be explained by (1) an increase in adenylyl cyclase activity or (2) a decrease in phosphodiesterase activity. This conclusion is based upon forskolin's activation of adenylyl cyclase without the interaction of G proteins (Daly, 1984). However, given that forskolin may also exert actions independent of adenylyl cyclase (Hoshi et al., 1988; Wagoner and Pallotta, 1988), a role of other post-receptor transduction elements in the supersensitivity cannot be eliminated at this time.

The differences between the right and left atria may be more quantitative than qualitative. The inotropic supersensitivity to isoprotenerol in left atria is about 5-fold, while that to forskolin is about 3-fold (Hawthorn et al., 1987). In right atria, the chronotropic and inotropic supersensitivity to isoproterenol are 2-fold (Torphy et al., 1982) and there is no indication of chronotropic supersensitivity to forskolin. Given the variability of cyclic AMP levels, it may be difficult to detect subtle changes in its generation associated with the small degree of supersensitivity characteristic of the right atrium.

There are differences in the pattern of supersensitivity in guinea-pig ventricular strips in comparison to atria. Tenner et al. (1988) reported supersensitivity to isoproterenol (3.3-fold); impromidine, a histamine H_2 receptor agonist (1.9-fold); and forskolin (3-fold) induced by pretreating guinea pigs with reserpine. Binding of [^{3}H]-dihydroalprenolol was increased

60 percent. Although an increase in density of beta adrenoceptors may contribute to the supersensitivity, the modest degree of the increase in binding and the supersensitivity to forskolin led Tenner et al. (1988) to conclude that there may also be a change in the transduction process.

Others have hypothesized that the G protein-adenylyl cyclase-cyclic AMP system is implicated in the development of cardiac supersensitivity in rats (Pik and Wollemann, 1977) and rabbits (Tkachuk and Wollemann, 1979). However, as discussed by Fleming (1984), the design of those experiments makes it difficult to relate the results to adaptive supersensitivity. Cros and McNeill (1987) demonstrated that pretreatment of guinea pigs with reserpine, 2.5 mg/kg for 2 days, selectively increased the slope and maximum response of the concentration-response curve for epinephrine's activation of adenylyl cyclase in membrane preparations from whole hearts. However, such doses of reserpine are likely to have direct effects on myocardial cells, independent of transmitter depletion (Fleming, 1984).

Acknowledgements

The authors appreciate the technical assistance of L. Bennett and W. Nadler. Partial support of this project came from NIH grant GM29840.

References

Cros GH, McNeill JH (1987) Reserpine-induced supersensitivity in adenylate cyclase preparations from guinea-pig heart. Eur J Pharmacol 139: 97–101

Crout JR, Muskus AJ, Trendelenburg U (1962) Effect of tyramine on isolated guinea-pig atria in relation to their noradrenaline stores. Br J Pharmacol 18: 600–611

Daly JW (1984) Forskolin, adenylate cyclase, and cell physiology: an overview. Adv Cyclic Nucl Res 17: 81–89

Fleming WW (1976) The variable sensitivity of excitable cells: possible mechanisms and biological significance. Rev Neurosci 2: 43–91

Fleming WW (1981) Postjunctional supersensitivity: a cellular homeostatic mechanism. Trends Pharmacol Sci: 152–154

Fleming WW (1984) A review of postjunctional supersensitivity in cardiac muscle. In: Fleming WW, Langer SZ, Graefe K-H, Weiner N (eds) Neuronal and extraneuronal events in autonomic pharmacology. Raven Press, New York, pp 205–219 (Proceedings of the symposium in honor of U Trendelenburg, Paris 1983)

Fleming WW, Trendelenburg U (1961) The development of supersensitivity to norepinephrine after pretreatment with reserpine. J Pharmacol Exp Ther 133: 41–51

Fleming WW, Westfall DP (1988) Adaptive supersensitivity. In: Trendelenburg U, Weiner N (eds) Catecholamines (Handbook Exp Pharmacol 90/I: 509–559)

Fleming WW, McPhillips JJ, Westfall DP (1973) Postjunctional supersensitivity and subsensitivity of excitable tissues to drugs. Rev Physiol Biochem Exp Pharmacol 68: 55–119

Goto K, Longhurst PA, Cassis LA, Head RJ, Taylor DA, Rice PJ, Fleming WW (1985) Surgical sympathectomy of the heart in rodents and its effect on sensitivity to agonists. J Pharmacol Exp Ther 234: 280–287

Hawthorn MH, Taylor DA, Fleming WW (1987) Characteristics of adaptive supersensitivity in the left atrium of the guinea pig. J Pharmacol Exp Ther 241: 453–457

Hoshi T, Garber SS, Aldrich RW (1988) Effect of forskolin on voltage-gated K^+ channels is independent of adenylate cyclase activation. Science 240: 1652–1658

Lowry OH, Rosenbrough NJ, Farr AL, Randall RJ (1951) Protein measurement with Folin reagents. J Biol Chem 193: 265–274

Pik K, Wollemann M (1977) Catecholamine hypersensitivity of adenylate cyclase after chemical denervation in rat heart. Biochem Pharmacol 26: 1448–1449

Rice PJ, Taylor DA, Valinsky WA, Head RJ, Fleming WW (1987) Norepinephrine depletion and sensitivity changes in rat heart induced by pretreatment with reserpine. J Pharmacol Exp Ther 240: 764–771

Schulz JC, Fleming WW, Westfall DP, Millechia R (1984) Cellular potentials, electrogenic sodium pumping and sensitivity in guinea-pig atria. J Pharmacol Exp Ther 231: 181–188

Taylor DA, Westfall DP, deMoraes S (1976) The effect of pretreatment with reserpine on the diastolic potential of guinea-pig atrial cells. Naunyn-Schmiedebergs Arch Pharmacol 293: 81–87

Tenner TE (1983) Propranolol withdrawal supersensitivity in rat cardiovascular tissue, in vitro. Eur J Pharmacol 92: 91–97

Tenner TE, Carrier O (1978) Reserpine-induced supersensitivity to the chronotropic and inotropic effects of calcium in rabbit atria. J Pharmacol Exp Ther 205: 185–192

Tenner TE, Young J, Riker J, Ramanadham S (1988) Nonspecific supersensitivity induced by reserpine in guinea pig cardiac ventricle tissue. J Pharmacol Exp Ther 246: 1–6

Tkachuk VA, Wollemann M (1979) Hypersensitivity to isoproterenol in rabbit heart decreases guanine nucleotide effect on adenylate cyclase. Biochem Pharmacol 28: 2097–2100

Torphy TJ, Westfall DP, Fleming WW (1982) Effect of reserpine pretreatment on mechanical responsiveness and [^{125}I] iodohydroxybenzylpindolol binding sites in the guinea pig right atrium. J Pharmacol Exp Ther 223: 332–341

Trendelenburg U, Gravenstein JS (1958) Effect of reserpine pretreatment on stimulation of the accelerans nerve of the dog. Science 128: 901–902

Wagoner PK, Pallotta BS (1988) Modulation of acetylcholine receptor desensitization by forskolin is independent of cAMP. Science 240: 1655–1657

Westfall DP, Fleming WW (1968a) Sensitivity changes in the dog heart to norepinephrine, calcium and aminophylline resulting from pretreatment with reserpine. J Pharmacol Exp Ther 159: 98–106

Westfall DP, Fleming WW (1968b) The sensitivity of the guinea pig pacemaker to norepinephrine and calcium after pretreatment with reserpine. J Pharmacol Exp Ther 164: 259–269

Authors' address: W. W. Fleming, Ph.D., Department of Pharmacology and Toxicology, West Virginia University Health Sciences Center, Morgantown, WV 26056, U.S.A.

J Neural Transm (1991) [Suppl] 34: 187–194

Molecular aspects of the receptor activation by imidazolines: an overview

P. N. Patil, D. R. Feller, and **D. D. Miller**

Ohio State University, College of Pharmacy, Divisions of Pharmacology and
Medicinal Chemistry, Lloyd M. Parks Hall, Columbus, Ohio, U.S.A.

Summary. Based on the pharmacological activity of chiral imidazolines the
steric requirements for the activation of the α-adrenoceptor are provided.
Importantly, the sequence of interaction of the critical groups of the asymmetric carbon of imidazolines and catecholamines with the α-adrenoceptor is
postulated. Thus, initial determinants of molecular efficacy are hypothesized.
The effect of aromatic fluoro-substitution and introduction of a double bond in
the imidazoline moiety on the pharmacologic activity is discussed. The unique
mechanism of *non*-adrenergic vascular activity of isothiocyanato-tolazoline is
presented.

Introduction

The class of directly acting naphazoline-like sympathomimetic imidazolines
were introduced in therapeutics as vasoconstrictor nasal decongestants (Sneader,
1985, U.S. patent #2161938). When compared with the catecholamines, the
long duration of the vascular response of the imidazolines is mainly mediated
by the activation of α-adrenoceptors. Several imidazolines however, behave as
partial agonists (Sanders et al., 1975; Ruffolo, 1983). The antihypertensive
action of clonidine, in part, is attributed to the prejunctional α_2-adrenoceptor-
mediated inhibition of the transmitter release. For imidazolines the ratio of
α_2/α_1 adrenoceptor-related affinities vary to a significant extent (Starke, 1981;
Sengupta et al., 1987). β-Adrenoceptor-mediated activities of imidazolines
were either weak or absent (Banning et al., 1984). Several imidazolines do not
appear to be substrates for neuronal catecholamine transport, catechol-O-
methyl-transferase and monoamine oxidase. Unlike catecholamines, the
liposolubility of imidazolines is much greater and expected to produce central
nervous system-mediated effects. Chemical structures of some imidazolines are
presented in Fig. 1.

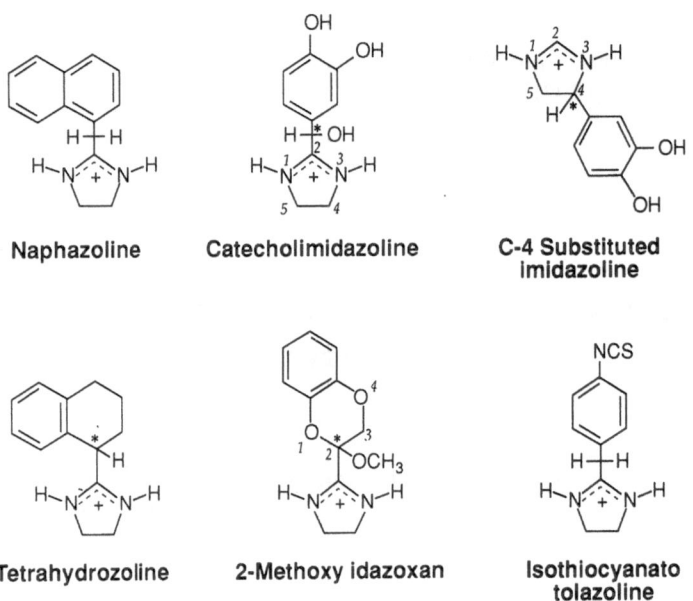

Fig. 1. Imidazoline derivatives

Steric requirements for the activation of the receptor

The receptor interactive conformation of naphazoline was postulated in which, the *naphthyl* and the *imidazoline* moieties were considered perpendicular with each other with a dihedral angle of ~90 degrees (Pullman et al., 1972). In such a conformation one nitrogen of the imidazoline moiety remains *trans* to the phenyl group. The predicted preferred conformation of catecholamines for the activation of adrenoceptors is similar in which the catechol moiety and the amino group remain *trans* to each other (reviewed by Patil et al., 1974). There is no doubt that the specific conformation of the reactive functional groups of the molecule is essential for the activation of the receptor; importantly, *the rate of change of the specific conformation of the receptor by the agonist in the physiologic millieu may be vital for the transduction.* Available methods do not reveal such information. Nonetheless, asymmetric imidazolines provide some interesting clues about steric aspects of the drug-receptor interaction and a *sequence* for the association of groups with the receptor.

The rank order of activation of prazosin-sensitive α_1-adrenoceptor of rat aorta was: desoxycatecholimidazoline > R(−)-catecholimidazoline > S(+)-catecholimidazoline. Analogues of catecholamidine or tolazoline follow a similar rank order of potency on a variety of tissues (Banning et al., 1984; Ahn et al., 1986; Rice et al., 1987; Sengupta et al., 1987). This rank order of the molecules for the activation of receptors was markedly different from that observed for desoxy-epinephrine and stereoisomers of epinephrine (Patil et al., 1974). The higher activity of desoxy-imidazolines over that of the R($-$) isomers indicated an interesting deviation from that of the Easson-Stedman postulate for the activation of the adrenoceptor. A benzylic hydroxyl group in the imidazoline series may cause a decrease in affinity because the activity of the desoxy form was consistently higher than the R($-$)-isomer. The molecular mode of interaction of the catecholimidazolines and the catecholamines with the α-adrenoceptor must differ (Banning et al., 1984).

The receptor-related affinity and the intrinsic efficacy are the important determinants of the potency of the agonist. The efficacy is connected with the early thermodynamics of the interaction. In the rat vas deferens, if the change in Gibb's free energy ($\Delta G°$) was calculated from the affinity constant generated for the α-adrenoceptor, values of -8.1 and -6.2 kcal/mol for R($-$)- and S($+$)epinephrine, respectively, were obtained. The interaction was largely driven by the entropy of the drug-receptor complex. About a 2 kcal/mol difference in the $\Delta G°$ values for stereoisomers of epinephrine was consistent with the energy required of hydrogen bond formation with the receptor. Similarly, about a 1 kcal/mol difference in the $\Delta G°$ values for the stereoisomers of catecholimidazoline indicates a weak hydrogen bond formation with the receptor but less favorably than that by R($-$)-epinephrine (Rice et al., 1989; Rice, 1983). In both types of molecules, namely catecholamines and imidazolines, the R sequence rule around the chiral center with benzylic hydroxyl group was maintained. Some amino acid residues, (Asp[113], Ser[204], Ser[207] and Phe[290]) of the receptor were considered vital for the interaction of adrenergic ligands (Strader et al., 1989).

Even though it is difficult to prove experimentally, in all probability initial determinants of efficacy may include a specific *sequence* of interaction of a few critical groups of the agonist with the active-site region of the adrenoceptor. For the α-adrenoceptor activation, the imidazoline moiety of tolazoline (or catecholimidazoline) may *first* interact with an anionic site, and the aromatic group then may bind through a π interaction. The sequence for the α-adrenoceptor stimulant action for R($-$)-norepinephrine may be that the protonated amino group (NH_3^+) takes the *first* priority, the benzyl-catechol the second and the benzylic hydroxyl group may be the last. For the activation of β-adrenoceptors, however, the benzylic hydroxyl group of R($-$)-norepinephrine may be the *first* anchoring group with the receptor sites and the NH_3^+ group may take the last priority. Only a few imidazolines were weak β-adrenoceptor activators and therefore, prediction about the sequence for the interaction is unnecessary.

As the result of the sequence of interaction of the groups, specific types of conformational changes in the receptor may occur. This may cause changes in cellular calcium and other biochemical events leading to the expression of efficacy. The proposed sequences of the functional groups for the receptor interaction may or may not follow the priority of the groups according to the Cahn-Ingold-Prelog sequence rule, in which higher atomic number precedes lower.

Asymmetry at other carbons

Methyl or benzyl group substitution at the carbon-4 (C-4) position of the imidazoline moiety of naphazoline reduced the affinity of the molecule with little or no difference in the dissociation constants of the blockers (Miller et al., 1976; Fuder et al., 1981). Introduction of similar asymmetry in catecholimidazolines, however, retained some intrinsic activity with little or no steric preference in the activation of α_1- or α_2-adrenoceptors (Rice et al., 1987; Sengupta et al., 1987). If however, the catechol moiety was attached directly at C-4, an interesting change in activity on rat aorta occurred; the S($-$)-isomer was about 20 times more potent than the R($+$)-isomer (unpublished data, 1990). Furthermore, if an extra carbon is added between the catechol moiety and C-4 of imidazoline, the α_1-adrenoceptor stimulatory activity of the R-isomer was nearly equal to the S-isomer (Miller et al., 1990). In human platelet aggregation studies, however, the S-isomer was 70 times more potent than the R-isomer. Why is the steric requirement for the activation of the adrenoceptor altered? For the activation of either α- or β-adrenoceptors, the enantiomers with 1R stereochemistry are more potent than their counterparts with 1S stereochemistry. Hydroxyl group substitution at the carbon bridge between the C-4 and catechol moiety is needed to provide the optimum absolute steric requirement for receptor activation.

The chiral center in tetrahydrozoline is unique. The characteristics of the dose-response curves of the stereoisomers of tetrahydrozoline indicated that the α-adrenoceptors of rabbit stomach and aorta may be different (Fuder et al., 1981). Enantiomers of idazoxan showed a few-fold difference in the affinity for [^{3}H]idaxozan interacting sites in rat cerebral cortex membranes. After substitution of a methoxy group at the chiral center, the affinity difference between the isomers was 640 fold; the R chirality of the molecule maintained higher affinity (Clark et al., 1986). These isomers should be useful to characterize subtypes of idazoxan specific receptors.

Halogen substituted catecholimidazolines

Substitution of fluorine (F) either at the 2, 5, or 6 position of the aromatic ring of the desoxycatechol-imidazolines retained α_1-adrenoceptor activating properties with the rank order of potency as follows: desoxy-catecholimidazoline $\geq$ 2F-desoxycatecholimidazoline $>$ 5F-desoxy-catecholimidazoline $>$ 6F-desoxycatecholimidazoline (Lamba Kanwal et al., 1988). The latter molecule was a partial agonist. This rank order of potency is different from corresponding F analogues of norepinephrine where 6-F norepinephrine is the potent α-agonist while 2F-norepinephrine is the potent β-agonist (Kirk and Creveling, 1984). The electronegativity of F produces a reactive positive charge around the aromatic moiety so that the receptor related conformation of the benzylic *hydroxyl* and *amino* group may be staggered. The fluoro catechol moiety and amino groups were postulated to exist in a *trans* conformation. The substitution of F in either the 6 or 2 position of the aromatic ring of norepinephrine can direct the two other groups into an α- or β-receptor active-conformation. Fluoro substituted imidazolines lack potent β-adrenoceptor activating properties, therefore, speculations advanced for norepinephrine analogues may not be simple extrapolated to imidazolines. Substitution of a bulky iodine atom in the aromatic ring of catecholimidazoline reduced adrenoceptor-mediated activity (Venkataraman et al., 1991).

Imidazolines and imidazoles

Introduction of double bonds between the C-4 and C-5 position of imidazoline produces a flat ring structure called imidaz*ole*. Naphaz*ole* has 1/15th the α$_1$-stimulant activity of naphaz*oline* (Lamba-Kanwal et al., 1988). As compared to catecholimidazoline the activity of catecholimidaz*ole* also decreased by one order of magnitude. pKa Values may drop from a high value (>9) in imidaz*olines* to about 7 for some imidaz*oles*. It is interesting to note that adenosine, histamine and pilocarpine can be viewed as derivatives of imidazoles. Do these molecules compete with imidazoline-receptors?

Adrenoceptor-mediated effects of imidazolines and the imidazoline preferring sites or receptors?

The rat vas deferens, desensitized to the stimulant effect of an imidazoline, remained responsive to the α-adrenoceptor activator phenylephrine, and provided the earliest clue to the fact that these two types of directly acting sympathomimetic drugs may act by different mechanisms (Ruffolo et al., 1977). On the rabbit aorta, nonphenolic imidazolines are partial agonists, and an additional response of the tissue can be achieved by epinephrine (Sanders et al., 1975). The slope of the concentration effect curve to (+)-hydroxytolazoline was different from that to the (−)-isomer. Prazosin blocked the response at the lower part of the curve and the prazosin-resistant component of the action persisted (Sengupta et al., 1987). Importantly, the isothiocyanato analogue of tolazoline produced a phentolamine-resistant response of rat aorta which was competitively blocked by idazoxan with an apparent $K_B = 7.7 \times 10^{-9}$ M (Venkataraman et al., 1989). The response to norepinephrine in the chlorethylclonidine-treated aortic tissue was not blocked by the α-adrenoceptor blockers (Oriowo and Bevan, 1990).

All these facts together with the studies on the subcellular fractions indicate that some imidazolines exhibit varying degreess of affinities for subtypes of adrenoreceptors (Han et al., 1987; Brown et al., 1990) and imidazoline preferring sites (Bousquet et al., 1984; Ernsberger et al., 1990;

Parini et al., 1989; Michel et al., 1990). Injection of imidazolines including clonidine in the rostral ventrolateral medulla produced hypotension which was correlated with the affinity at the imidazoline site and not with the affinity at the α_2-receptor (Ernsberger et al., 1990). It is interesting to note that histamine also interacts with the latter site.

Sympathetic nerve stimulation is reported to release *non*-adrenergic substances at the synapse. Therefore, the presence of various receptors at the post-synaptic membranes is highly probable. The relative distribution of various types of receptors at a given synapse is not known. The evidence, however; indicates that imidazolines may interact with different types of receptors. From the rat liver, an imidazoline interacting protein has been isolated (Parini et al., 1989) but neither the cellular distribution nor the functional correlates for such an interaction is known.

Partial agonist properties and the long duration of the response of imidazolines justify the use of these medications as nasal and ocular decongestants. Epinephrine is a potent vascular α_1-adrenoceptor activator and platelet aggregator. The latter activity of many potent vasoconstrictor imidazolines is weak or even absent. Furthermore, effects of imidazolines on nasal, lacrimal, gastric acid and salivary secretion remain to be investigated. New synthetic imidazolines and their stereoisomers in particular provide a precise tool to elucidate the pharmacologic characteristics of receptors.

Acknowledgements

The authors acknowledge support of this work by USPHS Grant No. GM-29358.

References

Ahn CH, Hamada A, Miller DD, Feller DR (1986) Alpha-adrenoceptor-mediated actions of optical isomers and desoxy analogs of catecholimidazoline and norepinephrine in human platelets: in vitro. Biochem Pharmacol 35: 4095–4102

Banning JW, Rice PJ, Miller DD, Ruffolo RR, Hamada A, Patil PN (1984) Differences in the adrenoceptor activation of stereoisomeric catecholimidazolines and catecholamines In: Fleming WW, et al (eds) Neuronal and extraneuronal events in autonomic pharmacology. Raven Press, New York, pp 167–180

Bousquet P, Feldman J, Schwartz J (1984) Central cardiovascular effects of alpha adrenergic drugs: differences between catecholamines and imidazolines. J Pharmacol Exp Ther 230: 232–236

Brown CM, Mackinnon AC, McGrath JC, Spedding M, Kilpatrick AT (1990) α_2-Adrenoceptor subtypes and imidazoline-like binding sites in the rat brain. Br J Pharmacol 99: 803–809

Clark RD, Michel AD, Whiting R (1986) Pharmacology and structure-activity relationships of α_2-adrenoceptor antagonists. In: Ellis G-P, West GB (eds) Progress in medicinal chemistry, vol 23. Elsevier, New York, pp 1–39

Ernsberger P, Giuliano R, Willette RN, Reis DJ (1990) Role of imidazole receptors in the vasodepressor response to clonidine analogs in the rostral ventrolateral medulla. J Pharmacol Exp Ther 253: 408–418

Fuder H, Nelson WL, Miller DD, Patil PN (1981) Alpha adrenoreceptors of rabbit aorta and stomach fundus. J Pharmacol Exp Ther 217: 1–9

Han C, Abel PW, Minneman KP (1987) Heterogeneity of α_1-adrenergic receptors revealed by chlorethylclonidine. Mol Pharmacol 32: 505–510

Kirk KL, Creveling CR (1984) The chemistry and biology of ring-fluorinated biogenic amines. Med Res Rev 4: 189–220

Lamba-Kanwal VK, Hamada A, Adejare A, Clark MT, Miller DD, Patil PN (1988) Activation of alpha-l adrenoreceptors of rat aorta by analogs of imidazoline. J Pharmacol Exp Ther 245: 793–797

Michel MC, Regan JW, Gerhardt MA, Neubig RR, Insel PA, Motulsky HJ (1990) Noradrenergic [^{3}H] idazoxan binding sites are physically distinct from α_2-adrenergic receptors. Mol Pharmacol 37: 65–68

Miller DD, Hsu FL, Ruffolo Jr RR, Patil PN (1976) Stereochemical studies of adrenergic drugs. Optically active derivatives of imidazolines. J Med Chem 19: 1382–1384

Miller DD, Hamada A, Clark MT, Adejare A, Patil PN, Shams G, Romstedt KJ, Kim SU, Instrasuksri U, Mckenzie JL, Feller DR (1990) Synthesis and α_2-adrenoceptor effects of substituted catecholimidazoline and catecholimidazole analogues in human platelets. J Med Chem 33: 1138–1144

Oriowo MA, Bevan JA (1990) Chlorethylclonidine unmasks a non-α-adrenoceptor noradrenaline binding site in the rat aorta. Eur J Pharmacol 178: 243–246

Parini A, Coupry I, Graham RM, Uzielli I, Atlas D, Lanier, SM (1989) Characterization of an imidazoline/guanidium receptive site distinct from the α_2-adrenergic receptor. J Biol Chem 264: 11874–11878

Patil PN, Miller DD, Trendelenburg U (1974) Molecular geometry and adrenergic drug activity. Pharmacol Rev 26: 323–392

Pullman B, Coubeils JL, Courriere PH, Gervois JP (1972) Quantum mechanical study of the conformational properties of phenethylamines of biochemical and medicinal interest. J Med Chem 15: 17–21

Rice PJ (1985) Molecular aspects of the activation of alpha-adrenergic receptors. Dissertation, The Ohio State University, Columbus, OH, U.S.A.

Rice PJ, Hamada A, Miller DD, Patil PN (1987) Asymmetric catecholimidazolines and catecholamidines: affinity and efficacy relations at the alpha adrenoreceptor in rat aorta. J Pharmacol Exp Ther 242: 121–130

Rice PJ, Miller DD, Sokoloski TD, Patil PN (1989) Pharmacologic implications of α-adrenoreceptor interactive parameters for epinephrine enantiomers in the rat vas deferens. Chirality 1: 14–19

Ruffolo RR (1983) Structure-activity relationship of alpha-adrenoceptor agonists In: Kunos G (ed) Adrenoceptors and catecholamines action. Wiley, New York, pp 1–50

Ruffolo RR, Turowski B, Patil PN (1977) Lack of cross desensitization between two structurally dissimilar α-adrenoceptor agonists. J Pharm Pharmacol 29: 378–380

Sanders J, Miller DD, Patil PN (1975) Alpha adrenergic and histaminergic effects of tolazoline-like imidazolines. J Pharmacol Exp Ther 195: 362–371

Sengupta JN, Hamada A, Miller DD, Patil PN (1987) Interaction of enantiomers of hydroxy tolazoline with adrenoceptors. Naunyn-Schmiedebergs Arch Pharmacol 335: 391–396

Sneader W (1985) Drug discovery: the evolution of modern medicines. Wiley, New York, p 105

Strader CD, Sigal IS, Dixon RAF (1989) Genetic approaches to the determination of structure function relationships of G protein-coupled receptors. TIPS [Suppl 26–30]

Starke K (1981) Presynaptic receptors. Ann Rev Pharmacol Toxicol 21: 7–30

Venkataraman BV, Hamada A, Shams G, Miller DD, Feller DR, Patil PN (1989)
    Paradoxical effects of isothiocyanate analog of tolazoline on rat aorta and human
    platelets. Blood Vessels 26: 335–346
Venkataraman BV, Sham G, Hamada A, Amemiya Y, Tantishaiyakul V, Hsu F,
    Fashempour J, Romstedt KJ, Miller DD, Feller DR, Patil PN (1991) Structure-
    activity studies of new imidazolines on adrenoceptors of rat aorta and human
    platelets. Naunyn-Schmiedebergs Arch Pharmacol 344: 454–463

Authors' address: P. N. Patil, Ph.D., College of Pharmacy (Parks Hall), 500 W.
12th Avenue, Columbus, Ohio 43210, U.S.A.

J Neural Transm (1991) [Suppl] 34: 195–201
© by Springer-Verlag 1991

5-HT$_4$-like receptors in mammalian atria

A. J. Kaumann

SmithKline Beecham Pharmaceuticals, Welwyn, and Clinical Pharmacology Unit, Department of Medicine, University of Cambridge, United Kingdom

Summary. Atrial myocardium of man and pig possess receptors that mediate positive inotropic effects and/or positive chronotropic effects of 5-hydroxytryptamine (5-HT). These 5-HT receptors are blocked with moderate affinity ($pK_B = 6.7–6.9$) by 3α-tropanyl-1H-indole-3-carboxylate (ICS 205930) but not by antagonists of α- and β-adrenoceptors, or 5-HT$_1$ subtypes, 5-HT$_2$ and 5-HT$_3$ receptors. In human and porcine atrium the receptors also mediate 5-HT-induced increases of both cyclic AMP levels and cyclic AMP-dependent protein kinase activity. Human and porcine atrial 5-HT receptors are also partially activated by the benzamides renzapride and cisapride albeit with lower potency and efficacy than 5-HT. The properties of these atrial 5-HT receptors resemble those of "so called" 5-HT$_4$ receptors, positively coupled to the adenylyl cyclase of mouse embyonic colliculi neurons. However, for colliculi 5-HT$_4$ receptors the benzamides have greater efficacy and ICS 205930 lower affinity than for atrial 5-HT receptors. The atrial 5-HT receptors of man and pig are, therefore, designated 5-HT$_4$-like receptors. Guinea-pig night atria appear to have a mixture of 5-HT$_3$- and 5-HT$_4$-like receptors involved in the mediation of positive chronotropic effects of 5-HT.

Perspective

The recent history of some mammalian atrial 5-HT receptors is summarised in Table 1.

Trendelenburg (1960) took the first systematic look at mammalian atrial 5-HT receptors. He studied sinoatrial receptors that mediate positive chronotropic effects of cat and guinea pig and receptors that mediate positive inotropic responses in rabbit atria. Using the classification of Gaddum and Picarelli (1957) as a framework, he proposed that feline atrial 5-HT receptors belonged to the D class because they were blocked by (+) lysergic acid diethylamide (LSD). This finding was later confirmed in both right and left feline atria where LSD is a competitive antagonist ($pK_B = 7.8$; Kaumann, unpublished experiments). As expected from D receptors, the effects of 5-HT were unsurmountably antagonised by dibenzyline (phenoxybenzamine) (Kaumann, 1983, 1985) but not at all by the 5-HT$_2$

Table 1. Mammalian atrial 5-HT receptors

Cat	Rabbit	Guinea pig	Rat	Piglet and pig	Man	Reference
D	M	Mixed				Trendelenburg (1960)
non-5-HT$_1$, non-5-HT$_2$						Kaumann (1983, 1985, 1986)
5-HT$_1$-like						Saxena et al. (1985)
	5-HT$_3$					Fozard (1984)
		5-HT$_3$? and 5-HT$_4$-like?				Kaumann (this paper)
			5-HT$_2$			Docherty (1988)
				5-HT$_4$-like		Kaumann (1990) Villalon et al. (1990) Kaumann et al. (1991b)
					5-HT$_4$-like	Kaumann et al. (1989a, 1990a,b, 1991a)

receptor (Peroutka and Snyder, 1979) antagonist ketanserin (van Nueten et al., 1981), ruling out 5-HT$_2$ receptors (Kaumann, 1983, 1985). The affinities of 5-HT and carboxamidotryptamine (5-CT), estimated for feline atria by receptor occlusion with phenoxylbenzamine, were found to be lower than expected for 5-HT$_1$ receptors (Kaumann, 1986). Based on the greater chronotropic potency of 5-CT as opposed to 5-HT in anaesthetised cats, Saxena et al. (1985) concluded that feline sinoatrial receptors were 5-HT$_1$-like (Bradley et al., 1986). 5-HT$_{1A}$ and 5-HT$_{1D}$ receptors do not appear to be involved in the effects of 5-HT on feline atria because the 5-HT$_{1A}$ agonist 8-hydroxy-2-(N, N-dipropylamino) tetralin (8-OH-DPAT) is only a weak marginal partial agonist and yohimbine is ineffective (Kaumann, 1983, 1985) at a concentration (2 μM) that saturates 5-HT$_{1D}$ receptors (Heuring and Peroutka, 1987).

Trendelenburg (1960) showed that the effects of 5-HT in rabbit atrium were blocked by morphine and greatly reduced through depletion of noradrenaline stores by reserpine pretreatment. He concluded that in rabbit the cardiostimulant effects of 5-HT are due to release of noradrenaline via M receptors. This mechanism was confirmed in the seventies by Fozard (see Fozard, 1989). Fozard's work led to the development of 5-HT$_3$ receptor (Bradley et al., 1986) antagonists such as MDL 72222 and ICS 205930

(Richardson et al., 1985). Fozard (1984) was able with MDL 72222 to block both the noradrenaline release and cardiostimulant effects of 5-HT in the rabbit heart. Thus, the cardiostimulant effects of 5-HT in rabbit heart are currently considered to be mediated by 5-HT$_3$ receptors located on sympathetic nerve terminals.

Rat sinoatrial tachycardia induced by 5-HT is antagonised by ketanserin suggesting an involvement of 5-HT$_2$ receptors (Docherty, 1988).

5-HT$_4$-like receptors

In contrast to the preceding three species, each possessing a different atrial 5-HT receptor, another three species appear to share similar atrial 5-HT receptors, namely 5-HT$_4$-like receptors (Table 1). Human right atrium was the first system for which 5-HT$_4$-like receptors were proposed (Kaumann et al., 1989a) and its characterisation is making progress (Kaumann et al., 1990a,b, 1991a). Subsequently, 5-HT-induced tachycardia (Villalon et al., 1990; Kaumann, 1990) and positive inotropic effects in left atria (Kaumann et al., 1991b) were described in pig and piglet.

How are 5-HT$_4$ receptors defined? These receptors were first described in mouse embryonic colliculi neurones as mediating stimulation of adenylyl cyclase by 5-HT (Dumuis et al., 1988). Benzamides, such as renzapride and cisapride are powerful agonists, as potent as, and more efficacious than, 5-HT (Dumuis et al., 1989). The effects of both 5-HT and the benzamides are competitively antagonised by ICS 205930 with a pK$_B$ of 6.0 but not by antagonists of 5-HT$_{1A}$, 5-HT$_{1B}$, 5-HT$_{1C}$, 5-HT$_{1D}$, 5-HT$_2$ and 5-HT$_3$ receptors (Dumuis et al., 1989).

The human atrial 5-HT receptors that mediate positive inotropic effects of 5-HT resemble 5-HT$_4$ receptors of mouse embryonic colliculi neurones in the following:

- they mediate an increase in cyclic AMP levels,
- they are stimulated by renzapride and cisapride, and with low potency by 5-CT,
- they are blocked by ICS 205930.

Despite these similarities human atrial 5-HT receptors differ in two quantitative aspects (Table 2):

- The affinity of ICS 205930 is somewhat lower for colliculi 5-HT$_4$ receptors than for human atrial 5-HT receptors,
- Renzapride and cisapride are only partial agonists, both with regard to cyclic AMP and the positive inotropism, and they are less potent than 5-HT in human atrium. Cisapride has also lower affinity and efficacy than renzapride. On the other hand on colliculi the two benzamides are equipotent and more efficacious than 5-HT.

In view of these quantitative differences, human atrial 5-HT receptors are termed 5-HT$_4$-like receptors (Kaumann, 1990; Kaumann et al., 1991a).

Table 2. Relative potencies (RP) and intrinsic activities (IA) of agonists and affinity estimates (-Log K_B, M = pK_B) for ICS 205930

	Mouse embryonic collicull[a]		Piglet sinoatrial node[b]		Human right atrium[c]	
ICS 205930	$pK_B = 6.0$		$pK_B = 6.9$		$pK_B = 6.7$	
Agonist	RP	IA	RP	IA	RP	IA
5-HT	1	1	1	1	1	1
5-MeOT	1	1	0.5[d]	0.9[d]	nd	nd
5-CT	0.003	1	0.006	0.9	0.003	1
Renzapride	0.8	1.3	0.4	0.7	0.13	0.6
Cisapride	1	1.4	0.25	0.4	<0.08	0.3

[a] Increase in cyclic AMP levels (Dumuis et al., 1988, 1989)
[b] Increase in sinoatrial beating rate (Kaumann, 1990)
[c] Increase in contractile force (Kaumann et al., 1990a, b, 1991a)
[d] Unpublished experiments (Kaumann)
nd not determined; *5-MeOT* 5-methoxytryptamine; *5-CT* 5-carboxamidotryptamine

Piglet right atrial 5-HT receptors that mediate tachycardia greatly resemble human right atrial 5-HT$_4$-like receptors (Table 2, Kaumann, 1990). Consistent with 5-HT$_4$-like receptors, 5-HT increases cyclic AMP levels in piglet left atria, ICS 205930 blocks the positive inotropic effects of 5-HT with a pK_B of 6.7 and renzapride is a partial agonist (Kaumann et al., 1991b).

Cyclic AMP-dependent protein kinase activity is stimulated by 5-HT through 5-HT$_4$-like receptors as expected from the elevated cyclic AMP levels caused by 5-HT in human right atrium (Kaumann et al., 1989a, 1990b). As expected for a partial agonist, renzapride is a less efficacious stimulant of the protein kinase than 5-HT and cisapride only causes marginal stimulation (Kaumann et al., 1990a, 1991a). As expected from stimulation of the cyclic AMP pathway, 5-HT hastens relaxation of human atrial muscle, presumably through phosphorylation of phospholamban and/ or troponin I by the protein kinase (Kaumann et al., 1990b). Again, renzapride is less effective than 5-HT in hastening relaxation and cisapride is ineffective (Kaumann et al., 1991a).

5-HT is less efficacious than physiological catecholamines (Kaumann et al., 1989b) or (−)-isoprenaline in inhancing atrial contractile force and cyclic AMP levels, and in stimulating cyclic AMP-dependent protein kinase (Kaumann et al., 1990b). However, under conditions of blockade of neuronal amine uptake and in right atria obtained from β-adrenoceptor-blocked (usually β$_1$) patients, 5-HT is five times more potent than (−)-noradrenaline (Kaumann et al., 1990b). Chronic β$_1$-adrenoceptor blockade augments the human right atrial inotropic responsiveness mediated through both β$_2$-adrenoceptors (Hall et al., 1990) and 5-HT$_4$-like receptors (Kaumann et al., 1990c) but not through β$_1$-adrenoceptors (Hall et al., 1990). The nature of his hyperresponsiveness apears to be due to an

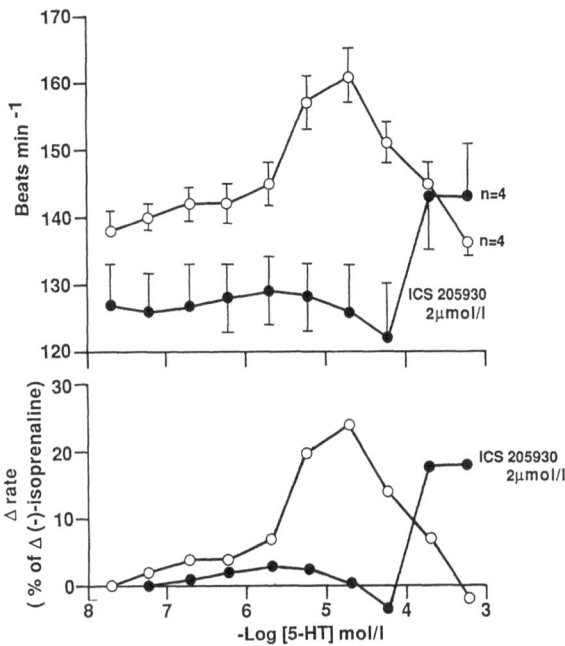

Fig. 1. Surmountable blockade of 5-HT-induced tachycardia by ICS 205930 in right atria obtained from reserpine-pretreated guinea pigs. Experimental conditions were those of Walter et al. (1984). The modified Krebs solution contained cocaine 6 μmol/l, (-)-bupranolol 1 μmol/l and ascorbate 200 μmol/l. A single concentration-effect curve to 5-HT was determined in each tissue. The experiment was terminated with (-)-isoprenaline (200 μmol/l). ICS 205930 was incubated for 1 h

enhanced coupling of β$_2$-adrenoceptors to effectors (adenylyl cyclase) by the G$_s$ protein system (Hall et al., 1990), but it is still unknown whether this also happens with 5-HT$_4$-like receptors. It should also be pointed out that 5-HT can also potently increase contractile force of left atrium isolated from hearts of patients with terminal heart failure and not treated with β-adrenoceptor antagonists. In these left atria ICS 205930 causes surmountable antagonism with a pK$_B$ = 6.8 (unpublished experiments by Louise Sanders and the author on atria from three patients), consistent with the existence of 5-HT$_4$-like receptors in human left atrium.

In guinea-pig right atria 2 receptor subtypes appear to mediate the positive chronotropic effects of 5-HT (Trendelenburg, 1960). The effects persist in reserpine-pretreated animals (Göthert and Klupp, 1978; Walter et al., 1984) ruling out noradrenaline release. The 5-HT responses can be antagonized by cocaine (Göthert and Klupp, 1978) but there are also cocaine-resistant responses (Walter et al., 1984). Cocaine is a weak 5-HT$_3$ receptor antagonist (Fozard, 1989) so that perhaps part of the effects of 5-HT could be mediated through 5-HT$_3$ receptors. The effects of 5-HT are also surmountably antagonized by ICS 205930 (Fig. 1) but to a somewhat greater extent than expected from its affinity for 5-HT$_4$ receptors (a pK$_B$ = 7.3 was estimated from the results of Fig. 1). ICS 205930 is a potent antagonist of 5-HT$_3$ receptors (Richardson et al., 1985) and a weak antagonist of 5-HT$_4$ receptors. The results of Fig. 1 are consistent with but do not prove the involvement of both 5-HT$_3$ and 5-HT$_4$ receptors. An unambiguous answer as to whether both or which 5-HT receptor participate(s) in the positive chronotropic effects of 5-HT in guinea-pig atria must await the arrival of a 5-HT$_4$-selective antagonist.

References

Bradley PB, Engel G, Feniuk W, Fozard JR, Humphrey PPA, Middlemiss DN, Mylecharane EJ, Richardson BP, Saxena PR (1986) Proposals for the classification and nomenclature of functional receptors for 5-hydroxytryptamine. Neuropharmacology 25: 563–576

Docherty JR (1988) Investigations of cardiovascular 5-hydroxytryptamine receptor subtypes in the rat. Naunyn-Schmiedebergs Arch Pharmacol 337: 1–8

Dumuis A, Buhelal R, Sebben M, Cory R, Bockaert J (1988) A non-classical 5-hydroxytryptamine receptor positively coupled with adenylate cyclase in the central nervous system. Mol Pharmacol 34: 880–887

Dumuis A, Sebben M, Bockaert J (1989) The gastro-intestinal prokinetic benzamide derivatives are agonists at the non-classical 5-HT receptor (5-HT$_4$) positively coupled to adenylate cyclase in neurones. Naunyn-Schmiedebergs Arch Pharmacol 340: 403–410

Fozard J (1984) MDL 72222, a potent and highly selective antagonist at neuronal 5-hydroxytryptamine receptors. Naunyn-Schmiedebergs Arch Pharmacol 326: 36–44

Fozard JR (1989) The development and early clinical evaluation of selective 5-HT$_3$ receptor antagonists. In: Fozard JR (ed) The peripheral actions of 5-hydroxytryptamine. Oxford University Press, Oxford, pp 354–376

Gaddum JH, Picarelli ZP (1957) Two kinds of tryptamine receptor. Br J Pharmacol 12: 323–328

Göthert M, Klupp N (1978) Cardiovascular effects of neurotoxic indolethylamines. Ann NY Acad Sci: 305: 457–477

Hall JA, Kaumann AJ, Brown MJ (1990) Selective β$_1$-adrenoceptor blockade enhances positive inotropic responses to endogenous catcholamines mediated through β$_2$-adrenoceptors in human atrial myocardium. Circ Res 66: 1610–1623

Heuring RE, Peroutka SJ (1987) Characteristics of a novel ^{3}H-5-hydroxytryptamine binding site subtype in bovine brain membranes. J Neurosci 7: 894–903

Kaumann AJ (1983) A classification of heart serotonin receptors. Naunyn-Schmiedebergs Arch 322: R42

Kaumann AJ (1985) Two classes of myocardial 5-hydroxytryptamine receptors that are neither 5-HT$_1$ nor 5-HT$_2$. J Cardiovasc Pharmacol 7 [Suppl 7]: S76–S78

Kaumann AJ (1986) Further differences between 5-HT receptors of atrium and venticle in cat heart. Br J Pharmacol 89: 546P

Kaumann AJ (1990) Piglet sinoatrial 5-HT receptors resemble human atrial 5-HT$_4$-like receptors. Naunyn-Schmiedebergs Arch Pharmacol 342: 619–622

Kaumann AJ, Sanders L, Brown AM, Murray KJ, Brown MJ (1989a) A receptor for 5-hydroxytryptamine in human atrium. Br J Pharmacol 98: 664P

Kaumann AJ, Hall JA, Murray KJ, Wells FC, Brown MJ (1989b) A comparison of the effects of adrenaline and noradrenaline on human heart: the role of β$_1$- and β$_2$-adrenoceptors in the stimulation of adenylate cyclase and contractile force. Eur Heart J 10 [Suppl B]: 29–37

Kaumann AJ, Sanders L, Brown AM, Murray KJ, Brown MJ (1990a) Human atrial 5-

HT receptors: similarity to rodent neuronal 5-HT$_4$ receptors. Br J Pharmacol 100: 319P

Kaumann AJ, Sanders L, Brown AM, Murray KJ, Brown MJ (1990b) A 5-hydroxytryptamine receptor in human atrium. Br J Pharmacol 100: 879–885

Kaumann AJ, Sanders L, Brown MJ (1990c) Chronic β-adrenoceptor blockade enhances positive inotropic responses to 5-hydroxytryptamine in human atrium. J Mol Cell Cardiol 22 [Suppl 111]: S2

Kaumann AJ, Sanders L, Brown AM, Murray KJ, Brown MJ (1991a) A 5-HT$_4$-like receptor in human right atrium. Naunyn-Schmiedebergs Arch Pharmacol 344: 150–159

Kaumann AJ, Brown AM, Raval P (1991b) Putative 5-HT$_4$-like receptors in piglet left atrium. Br J Pharmacol 102: 98 P

Peroutka SJ, Snyder SH (1979) Multiple serotonin receptors: differential binding of ^{3}H 5-hydroxytrptamine, ^{3}H lysergic acid diethylamide and ^{3}H spiperidol. Mol Pharmacol 16: 687–699

Richardson BP, Engel G, Donatsch P, Stadler PA (1985) Identification of serotonin M-receptor subtypes and their specific blockade by a new class of drugs. Nature 316: 126–131

Saxena PR, Mylecharane EJ, Heiligers J (1985) Analysis of the heart rate effects of 5-hydroxytryptamine in the cat; mediation of tachycardia by 5-HT$_1$-like receptors. Naunyn-Schmiedebergs Arch Pharmacol 330: 121–129

Trendelenburg U (1960) The action of histamine and 5-hydroxytryptamine on isolated mammalian atria. J Pharmacol Exp Ther 130: 450–460

Van Nueten JM, Janssen PAJ, Van Beck J, Xhonneux R, Verbeuren TJ, Vanhoutte PM (1981) Vascular effect of ketanserin (R 41468), a novel antagonist of 5-HT$_2$ serotonergic receptors. J Pharmacol Exp Ther 218: 217–230

Villalon CM, den Boer MO, Heiligers JPC, Saxena PR (1990) Mediation of 5-hydroxytryptamine-induced tachycardia in the pig by the putative 5-HT$_4$ receptor. Br J Pharmacol 100: 665–667

Walter M, Lemoine H, Kaumann AJ (1984) Stimulant and blocking effects of optical isomers of pindolol on the sinoatrial node and trachea of guinea pig. Role of β-adrenoceptor subtypes in the dissociation between blockade and stimulation. Naunyn-Schmiedebergs Arch Pharmacol 327: 159–175

Author's address: Dr. A. J. Kaumann, SmithKline Beecham Pharmaceuticals, The Frythe, Welwyn, Hertfordshire AL6 9AR, United Kingdom

J Neural Transm (1991) [Suppl] 34: 203–210

Glutamate receptor antagonism: neurotoxicity, anti-akinetic effects, and psychosis

P. Riederer[1], **K. W. Lange**[1], **J. Kornhuber**[1], and **K. Jellinger**[2]

[1]Clinical Neurochemistry, Department of Psychiatry, University of Würzburg,
Federal Republic of Germany
[2]Ludwig Boltzmann Institute of Clinical Neurobiology, Lainz Hospital,
Vienna, Austria

Summary. There is evidence to suggest that glutamate and other excitatory amino acids play an important role in the regulation of neuronal excitation. Glutamate receptor stimulation leads to a non-physiological increase of intracellular free Ca^{2+}. Disturbed Ca^{2+} homeostasis and subsequent radical formation may be decisive factors in the pathogenesis of neurodegenerative diseases.

Decreased glutamatergic activity appears to contribute to paranoid hallucinatory psychosis in schizophrenia and pharmacotoxic psychosis in Parkinson's disease. It has been suggested that a loss of glutamatergic function causes dopaminergic over-activity. Imbalances of glutamatergic and dopaminergic systems in different brain regions may result in anti-akinetic effects or the occurrence of psychosis. The simplified hypothesis of a glutamatergic-dopaminergic (im)-balance may lead to a better understanding of motor behaviour and psychosis.

Introduction

It is only recently that excitatory amino acid receptors have been discovered. Through the use of selective agonists and antagonists it has become evident that these receptors consist of different subtypes (for review see Watkins et al., 1990). At present the most useful classification provides the following excitatory amino acid receptor subtypes: N-methyl-D-aspartate (NMDA) receptors, kainate receptors, quisqualate receptors or α-amino-3-hydroxy-5-methyl-4-isoxazolepropionate (AMPA) receptors, metabotropic receptors and L-aminophosphonobutyrate (L-AP4) receptors. Some important features of these excitatory amino acid receptor subtypes are given in Table 1. It is evident that these receptors fulfill a variety of different physiological functions depending on their regional and subregional location, their pre- or postsynaptic localization, the neurotransmitter system at which they are located and their quantitative distribution.

Table 1. Glutamate receptor subtypes: major pharmacological profile and localization

Glutamate receptor subtype	Agonist	Antagonist	Modulating system	Ionic channel	Localization
NMDA	NMDA, L-aspartate, L-glutamate	PCP, Ketamine, MK-801, SKF-10047, Mg^{++} (N.C.), CPP, D-AP5, CGS 19755, DAA (C.)	polyamines, glycine +, D-serine +, MNQX −, 7-chlorokynurenate, HA 966	Na^+, K^+, Ca^{2+}	postsynaptic
AMPA	quisqualate, AMPA, glutamate	CNQX, NBQX, DGG, GDEE, babiturates, philanthotoxin		Na^+, Ka^+	postsynaptic, glial
Kainate	kainate, domoate				presynaptic
L-AP4	L-AP4, L-serine-O-phosphate				presynaptic
Metabotropic	quisqualate, ibotenate, ACPD	L-AP3, L-AP4		NO coupling to PLC-system and via IP_3 influences intracellular Ca^{2+} stores	postsynaptic, glial

Abbreviations: C. competitive; *N.C.* non-competitive; *ACPD* 1-amino-cyclopentane-1, 3-dicarboxylic acid; *AMPA* α-amino-3-hydroxy-5-methyl-4-isoxazolepropionic acid; *D-AP5* D-2-amino-5-phosphonopentanoic acid; *L-AP3* L-amino-3-phosphonopropionic acid; *L-AP4* L-2-amino-4-phosphonobutanoic acid; *CGS 19755* 4-phosphonomethyl-2-piperidinecarboxylic acid; *CNQX* 6-cyano-7-nitroquinoxaline-2, 3-dione; *CPP* (±)-2-carboxypiperazine-4-yl-propyl-1-phosphonic acid; *DAA* D-aminoadipate; *DGG* D-glutamylglycine; *GDEE* glutamic acid diethyl ester; *HA 966* 1-hydroxy-3-amino-pyrrolidin-2-one; *MK801* (+)-5-methyl-10, 11-dihydro-5H-dibenzo[a,d]cyclohepten-5,10-imine maleate; *MNQX* 6,8-dinitroquinoxalinedione; *NBQX* 6-nitro-7-sulphamobenzo[f]quinoxaline-2,3-dione; *NMDA* N-methyl-D-aspartate; *NO* nitric oxide; *PCP* phencyclidine; *SKF-10047* N-allyl-normetazocine; + activating; − inhibiting

NMDA receptors are involved in the regulation of intracellular free Ca^{2+}-concentrations. It has therefore been suggested that glutamate and other excitatory amino acids play an important role in the regulation of neuronal excitation and may bring about neuronal destruction if administered in sufficient excess.

Glutamate toxicity and neurodegeneration

The regulation of Ca^{2+} and its compartmentalization is important for both presynaptic and postsynaptic events. Alterations in the extracellular/intracellular Ca^{2+} ratio can produce deleterious changes in cell function.

Altered Ca^{2+} homeostasis may influence a variety of physiological cell functions including Ca^{2+} transport systems, Ca^{2+} binding proteins and Ca^{2+}-activated proteases (Gibson and Peterson, 1987). A non-physiological increase of intracellular free Ca^{2+} leads therefore to a dysregulation of membrane-dependent processes. This is generally accompanied by a loss of energy supply to the cell, and the loss of ATP in particular alters the biochemical homeostasis of the cell.

Neurodegeneration is caused or accompanied by Ca^{2+} influx and intracellular Ca^{2+} mobilization. It has been suggested that excitatory amino acids play an important role in excitatory neurotoxicity and neurodegeneration (for review see Meldrum and Garthwaite, 1990). Glutamate with its high concentration in the mammalian brain is the probable neurotransmitter at most excitatory synapses and the most likely excitatory amino acid toxin. NMDA receptor channels permit a large Ca^{2+} influx and there is evidence to suggest that toxic Ca^{2+} entry occurs mainly through these channels (Garthwaite and Garthwaite, 1987). The activation of proteases and other Ca^{2+}-dependent enzymes such as protein kinase C, phospholipases and Ca^{2+}-calmodulin-dependent protein kinase II may contribute to glutamate toxicity (Meldrum and Garthwaite, 1990). The substantial increase of Ca^{2+}-activated proteases such as calpains causes destruction of microtubules, neurofilaments, etc. and may induce derangement of structural membrane integrity. Calpains also convert xanthine dehydrogenase to xanthine oxidase and free radicals are subsequently generated during purine metabolism. Increased phospholipase activity results in the release of lipids and leads to production of arachidonic acid, which can be metabolized to produce free radicals. Radicals increase lipid peroxidation of membrane constituents and enhance the release of excitatory amino acids. In addition, arachidonic acid blocks the uptake of glutamate into glial and neuronal cells.

All of these pathological events cause a catabolic state in which nutritional supply decreases. Neuronal processes are eventually destroyed and neurodegeneration becomes uncontrolled and progressive. In Parkinson's disease and in Alzheimer's dementia a loss of about 70% of cell bodies in the substantia nigra pars compacta and the nucleus basalis Meynert, respectively, is necessary before the major clinical symptoms are observed. Whether or not disturbed Ca^{2+} homeostasis and radical formation are decisive pathobiochemical factors in these disorders is the subject of intensive research.

There is experimental evidence to suggest that excitotoxic mechanisms contribute to neuronal loss occurring as the result of cerebral ischaemia (for review see Meldrum and Garthwaite, 1990). Current knowledge suggests that a loss of glutamatergic function is a plausible hypothesis for the occurrence of productive symptoms in schizophrenia (see below). Decreased glutamatergic function may in theory be accompanied by a reduced rate of cerebral infarction in patients with schizophrenia. In order to examine this proposition we took an unselected series of 880 patients with neurological or psychiatric diseases who died at Lainz Geriatric Hospital, Vienna, between 1981 and 1988. In addition to routine autopsy, examinations were performed by a neuropathologist (K. J.). The number and percentage of patients dying of cerebral infarction was determined for the various diagnoses (Table 2). The number of such deaths was highest in the group of patients with a history of cerebral infarction. In contrast to neurological diseases, psychiatric disorders showed the lowest death rate caused by cerebral infarction. This preliminary evaluation is in line with the theories of decreased glutamatergic activity in schizophrenia and enhanced excitatory amino acid release in patients with cerebral infarction. It is interesting to note that cerebral infarction was not found in a group of 40 depressed patients. It is not known whether this can be directly related to decreased glutamatergic activity or to antidepressant therapy which may antagonize NMDA receptor function (Reynolds and Miller, 1988).

Table 2. Neurological and psychiatric patients dying of neuropathologically confirmed cerebral infarction (1981–1988)

Disease	Patients (n)	Sex F/M	Age (n) <65 >65 years		Presumed NMDA receptor density	Cerebral infarction as cause of death N %	
Cerebral infarction	77	48/29	13	64	↑ ↓	21	27
Dementia + cerebrovascular insufficiency	522	313/209	58	464	↑ = ↓	52	10
Parkinson's disease	34	21/13	4	30	↑ ↓	3	11
Other neurol. disorders (Huntington's disease, multiple sclerosis)	45	18/27	25	20	↑	9	20
Dementia of Alzheimer type	64	6/58	45	19	↑ ↓	3	5
Schizophrenia	105	66/39	45	60	↓	1	1
Depression	40	29/11	7	33	?	0	0

NMDA receptor densities are presumed to be increased (↑), decreased (↓) or unaltered (=) in comparison with control subjects. Increased, decreased or unaltered pathway activity may occur in the same disease depending on the loop systems involved. The quantity of change, however, is dependent on the progress and duration of the disease. This information is not given here

The role of glutamate in schizophrenia

It has recently been postulated that decreased glutamatergic function is a pathobiochemical marker of schizophrenia (Kim et al., 1980). A decreased release of glutamate has been found in the frontal and temporal cortex of schizophrenic patients (Sherman et al., 1991) while increased NMDA receptor density has been measured in the temporal and parietal cortex (Suga et al., 1990). In the putamen, increased (Kornhuber et al., 1989) and unaltered (Suga et al., 1990; Weissman et al., 1991) NMDA receptor densities have been reported. Quisqualate receptors are not changed in the frontal, temporal and parietal cortex (Kurumaji et al., 1990) while kainate receptor binding is increased in the frontal cortex (Deakin et al., 1989; Nishikawa et al., 1983) and not changed (Deakin et al., 1989) or decreased (Kerwin et al., 1988; Harrison et al., 1991) in the hippocampus. Taken together, the data available is of value only as a starting point for further research since both the number of studies and the number of brain regions examined are limited. For example, the most vulnerable brain regions in schizophrenia, the entorhinal cortex (Jakob and Beckmann, 1986) and the prefrontal cortex (Benes et al., 1986), have not been studied in detail and only preliminary biochemical evidence exists to suggest that NMDA receptor density is marginally increased in the entorhinal cortex (Kornhuber et al., 1989).

We assume that it is the loss of glutamatergic activity that induces an enhanced dopaminergic tone. The "dopamine hypothesis" of schizophrenia suggesting dopaminergic overactivity in the pathobiochemistry of some productive symptoms (paranoid hallucinatory psychosis) seems to be valid according to this assumption (Kornhuber et al., 1990).

Anti-akinetic effects of glutamate antagonists and pharmacotoxic psychosis

In Parkinson's disease enhanced glutamatergic activity is assumed to occur in the nucleus subthalamicus due to a decreased GABAergic input from the lateral globus pallidus. The cortico-striatal fibres also appear to be functionally over-active as the result of decreased dopaminergic nigrostriatal activity. By contrast, the glutamatergic thalamo-cortical pathway shows reduced activity due to GABAergic influence on the ventrolateral thalamus (Riederer and Berger, 1991).

The only anti-glutamatergic drugs available for the treatment of Parkinson's disease are the non-competitive NMDA receptor antagonists amantadine and memantine, which have only moderate anti-akinetic efficacy compared to dopamimetic substances (Schwab et al., 1969). However, threshold doses of memantine producing mild anti-akinetic effects result in pharmacotoxic psychosis in an unexpectedly high proportion of patients (Riederer et al., 1991). Amantadine is known to have anti-parkinsonian effects and pharmacotoxic psychoses are frequent adverse reactions (Danielczyk, 1973). In Parkinson's disease there is a lack of data

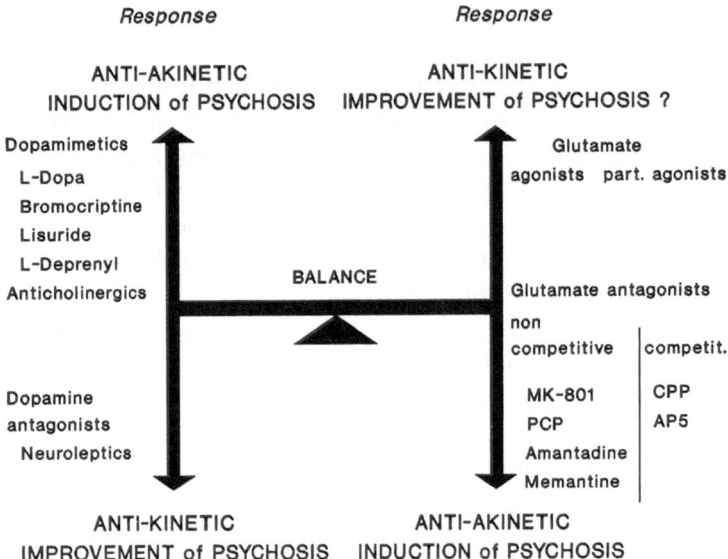

Fig. 1. Simplified illustration of glutamatergic-dopaminergic imbalances: anti-akinetic effects and psychosis

confirming a disturbance of glutamatergic function in limbic and cortical areas and supporting a glutamatergic hypothesis of pharmacotoxic psychosis. However, the fact that memantine has a considerable potential to induce pharmacotoxic psychosis at threshold doses which produce minor anti-akinetic effects, could suggest that glutamatergic activity in areas responsible for psychosis is reduced. Since under-active glutamatergic systems are further inhibited by NMDA receptor antagonists, adverse reactions such as pharmacotoxic psychosis are more likely to occur.

It is well known that all dopamimetic substances cause pharmacotoxic psychosis in Parkinson's disease and are able to aggravate productive symptoms in schizophrenia. It is not known, however, whether competitive NMDA receptor antagonists, which are known to enhance locomotor activity in experimental animals (Svensson et al., 1991; Löschmann et al., 1991), have potent anti-akinetic efficacy in Parkinson's disease or whether these substances also create the adverse reactions of dopamimetics and non-competitive NMDA receptor antagonists. The development of competitive glutamatergic antagonists or of partial agonists/antagonists could be another strategy capable of producing anti-akinetic effects with only mild side-effects. A simplified summary of this hypothesis is given in Fig. 1.

References

Benes FM, Davidson J, Bird ED (1986) Quantitative cytoarchitectural studies of the cerebral cortex of schizophrenics. Arch Gen Psychiatry 43: 31–35

Danielczyk W (1973) Die Behandlung von akinetischen Krisen. Med Welt 24: 1278

Deakin JWF, Slater P, Simpson MDC, Gilchrist AC, Skan WJ, Royston MC, Reynolds GP, Cross AJ (1989) Frontal cortical and left temporal glutamatergic dysfunction in schizophrenia. J Neurochem 52: 1781–1786

Garthwaite G, Garthwaite J (1987) Receptor-linked ionic channels mediate N-methyl-D-aspartate neurotoxicity in rat cerebellar slices. Neurosci Lett 83: 241–246

Gibson GE, Peterson C (1987) Calcium and the aging nervous system. Neurobiol Aging 8: 329–343

Harrison PJ, McLaughlin D, Kerwin RW (1991) Decreased hippocampal expression of a glutamate receptor gene in schizophrenia. Lancet i: 450–452

Jakob H, Beckmann H (1986) Prenatal developmental disturbances in the limbic allocortex in schizophrenics. J Neural Transm 65: 303–326

Kerwin RW, Patel S, Meldrum BS, Czudek C, Reynolds GP (1988) Asymmetrical loss of glutamate receptor subtype in left hippocampus in schizophrenia. Lancet i: 583–584

Kim JS, Kornhuber HH, Schmid-Burgk W, Holzmüller B (1980) Low cerebrospinal fluid glutamate in schizophrenic patients and a new hypothesis on schizophrenia. Neurosci Lett 20: 379–382

Kornhuber J, Mack-Burkhardt F, Riederer P, Hebenstreit GF, Reynolds GP, Andrews HB, Beckmann H (1989) [^{3}H]MK-801 binding sites in postmortem brain regions of schizophrenic patients. J Neural Transm 77: 231–236

Kornhuber J, Riederer P, Beckmann H (1990) The dopaminergic and glutamatergic systems in schizophrenia. In: Bunney WE, Hippius H, Laakmann G, Schmauß M (eds) Neuropsychopharmacology. Springer, Berlin Heidelberg New York Tokyo, pp 714–720

Kurumaji A, Ishimaru M, Toru M (1990) Quisqualate receptors in post-mortem brain of chronic schizophrenics. Proc Kyoto: New Trends in Schizophrenia and Mood Disorders Research, p 29

Löschmann PA, Lange KW, Kunow M, Rettig KJ, Jähnig P, Honore T, Turski L, Wachtel H, Jenner P, Marsden CD (1991) Synergism of the AMPA-antagonist NBQX and the NMDA-antagonist CPP with L-Dopa in models of Parkinson's disease. J Neural Transm [PD-Sect] 3: 203–213

Meldrum B, Garthwaite J (1990) Excitatory amino acid neurotoxicity and neurodegenerative disease. Trends Pharmacol Sci 11: 379–387

Nishikawa T, Takashima M, Toru M (1983) Increased ^{3}H-kainic acid binding in the prefrontal cortex in schizophrenia. Neurosci Lett 40: 245–250

Reynolds IR, Miller RJ (1988) Tricyclic antidepressants block N-methyl-D-aspartate receptors: similarities to the action of zinc. Br J Pharmacol 95: 95–102

Riederer P, Berger W (1991) Locomotion and behaviour: the interaction of loops and transmitter. Proc 5th World Congress of Psychiatry, Florence (in press)

Riederer P, Kornhuber J, Gerlach M, Danielczyk W, Youdim MBH (1991) Glutamatergic-dopaminergic imbalance in Parkinson's disease and paranoid hallucinatory psychosis. Proc Int Workshop on Parkinson's Disease, Berlin. Medicom Europe BV (in press)

Schwab RS, England AC, Poskanzer DC, Young RR (1969) Amantadine in the treatment of Parkinson's disease. J Am Med Assoc 208: 1168

Sherman AD, Davidson AT, Baruah S, Hegwood TS, Waziri R (1991) Evidence of glutamatergic deficiency in schizophrenia. Neurosci Lett 121: 77–80

Suga I, Kobayashi T, Ogata H, Toru M (1990) Increased ^{3}H-MK801 binding sites in post-mortem brains of chronic schizophrenic patients. Proc Kyoto: New Trends in Schizophrenia and Mood Disorders Research, p 28

Svensson A, Pileblad E, Carlsson M (1991) A comparison between the non-competitive NMDA antagonist dizocilpine (MK-801) and the competitive NMDA antagonist D-CPPene with regard to dopamine turnover and locomotor-stimulatory properties in mice. J Neural Transm [Gen Sect] 85: 117–129

Watkins JC, Kroogsgaard-Larsen P, Honore T (1990) Structure-activity relationships in the development of excitatory amino acid receptor agonists and competitive antagonists. Trends Pharmacol Sci 11: 25–33

Weissman AD, Casanova MF, Kleinman JE, London ED, DeSouza EB (1991) Selective loss of cerebral cortical sigma, but not PCP binding sites in schizophrenia. Biol Psychiatry 29: 41–54

Authors' address: Prof. Dr. P. Riederer, Clinical Neurochemistry, Department of Psychiatry, University of Würzburg, Füchsleinstrasse 15, D-W-8700 Würzburg, Federal Republic of Germany

J Neural Transm (1991) [Suppl] 34: 211–216

Stereotypy and asymmetry in mice

P. B. Dews

Laboratory of Psychobiology, Harvard Medical School, Boston, MA, and Behavioral Biology, New England Regional Primate Research Center, Harvard Medical School, Southborough, MA, U.S.A.

Summary. Mice ran in a circular runway. Number and direction of circuits were recorded. Most control mice ran about the same number of circuits in each direction. After 100 μmol/kg cocaine there were 3.5 times as many circuits and most mice ran most circuits in one direction. Some mice ran strongly in one direction after a first dose of cocaine and strongly in the other direction after a second dose. Hence, the primary influence toward unidirectional running is stereotypy rather than asymmetry.

Introduction

Amphetamine and related drugs cause many species of laboratory animals to perform one particular behavioral activity, such as sniffing in rodents, to the partial or complete exclusion of other behavioral activities (Randrup et al., 1963). The phenomenon was called stereotypy and has been extensively studied in the intervening years since it was reported (Cooper and Dourish, 1990).

Ungerstedt (1971a) reported that unilateral lesions in the basal ganglia of rats produced by injection of 6-hydroxydopamine caused the rat to rotate. The activity of 6-hydroxydopamine focussed attention on catecholamine systems, notably the dopamine system. It was found that amphetamine would enhance rotation in animals with lesions (Ungerstedt, 1971b). It was then found, however, that amphetamine could induce rotation in unlesioned, normal rats (Jerussi and Glick, 1974). The effect in normal animals was taken as evidence of brain asymmetry even in unlesioned animals (Glick et al., 1977). A similar effect was found with cocaine (Glick et al., 1983).

We found the mice in a circular runway would run around roughly as often in clockwise (Dir C) and counterclockwise (Dir CC) directions under control conditions (Dews, 1990). After 100 μmol/kg cocaine, however, not only did the number of circuits increase several fold, but there was a clear tendency for the circuits to be made more in a single direction (Dews, 1990). The tendency to unidirectional running could be the result of asymmetry in the brain. It could be, however, the result of stereotypy; the subject starts to run and simply continues running in that direction. In earlier studies along the same lines, Kokkinidis (1987) recognized that both asymmetry and stereotypy could be involved in more unidirectional turning or running after amphetamine. He concluded that stereotypy was the more important influence. Our results in the present experiments support his conclusion; while, at the same time, confirming asymmetry.

Material and methods

Results are based on 249 mice of Charles River CD-1 ancestry, 127 males and 122 females. Average ages were between 70 and 90 days and average weights were 29 g for males and 23 g for females.

Apparatus

Mice ran in a circular runway, a complete circuit being 1 m. Three beams of light on to photodiodes traversed the runway so that they were broken when the mouse passed. Thus frequency and direction of circuits could be recorded.

Schedule

Individual mice were placed in the darkened runway. Five minutes later the beams of light came on and circuits were tallied for 1,000 s.

Drug

Cocaine HCl was dissolved in 9 g/l saline so that an i.p. dose of 100 µmol/kg was contained in 0.01 ml/g. Each mouse received cocaine at least 2 times.

Results

Under control conditions the mice completed an average of 46 circuits and the average proportion in Dir C was 0.48. The frequency distribution of the proportion of circuits in Dir C was symmetrical, showing no bias toward Dir C or Dir CC, and was closely fit by a normal distribution curve (not shown) (Fig. 1). After 100 µmol/kg cocaine the average number of circuits was 173, almost 3.5 times greater than control and the average proportion in Dir C was 0.48, virtually the same as under control conditions. The frequency distribution of proportions in Dir C, however, was changed grossly. Whereas under control conditions, most mice made roughly the same number of circuits in Dir C and in Dir CC, after cocaine most mice made most circuits

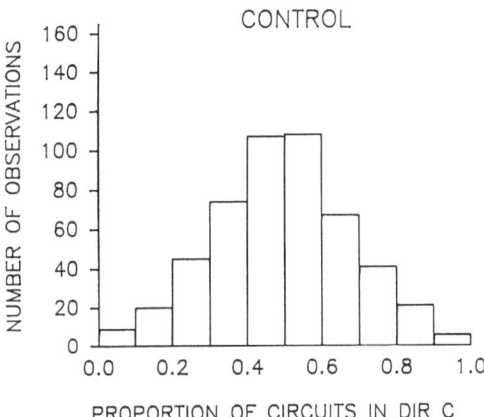

Fig. 1. Frequency distribution of proportion of circuits in Dir C in duplicate determinations in 249 mice under control conditions

either in Dir C or in Dir CC and few made even approximately equal numbers in the two directions (Fig. 2).

The frequency distribution of proportions of circuits in Dir C after the 2nd dose of cocaine was analysed separately for mice that had made a proportion of less than 0.5 and for those that made more than 0.5 of their circuits in Dir C after the 1st dose of cocaine (Fig. 2). Among the mice that had made a proportion in Dir C less than 0.5 after the 1st dose of cocaine, the distribution after the 2nd dose was skewed to the left i.e. most mice also had a proportion of less than 0.5 also after the 2nd dose. The most frequent class interval was less than 0.1 in Dir C. But no fewer than 30 of the 125 mice that made fewer circuits in Dir C after the 1st dose of cocaine made more in Dir C after the 2nd dose of cocaine. Of these, 14 made a proportion of greater than 0.9 in Dir C and 8 of these made all circuits in Dir C. Three of the mice that made all circuits in Dir C after the 2nd dose of cocaine had made 0 circuits in Dir C after the 1st dose of cocaine.

Similarly, of the 124 mice that had made a proportion of more than 0.5 circuits in Dir C after the 1st dose of cocaine most mice had a proportion greater than 0.5 after the 2nd dose of cocaine (Fig. 2). But no fewer than 35 mice of the 124 mice making more circuits in Dir C after the 1st dose of cocaine made fewer in Dir C after the 2nd dose of cocaine. Of these, 17 made a proportion less than 0.1 in Dir C and 9 made 0 circuits in Dir C. Five of the mice that made no circuits in Dir C after the 2nd dose of cocaine had made all circuits in Dir C after the 1st dose of cocaine.

There is a statistically highly significant tendency for mice that made a higher proportion of circuits in Dir C than in Dir CC after the 1st dose of cocaine to make a higher proportion of circuits in Dir C after the 2nd dose; and for mice that made a smaller proportion of circuits in Dir C after the 1st dose of cocaine to make a smaller proportion also after the 2nd dose (Table 1). There were 8 mice that made all circuits in a single direction after both doses of cocaine but in opposite directions after the 1st and 2nd doses. A

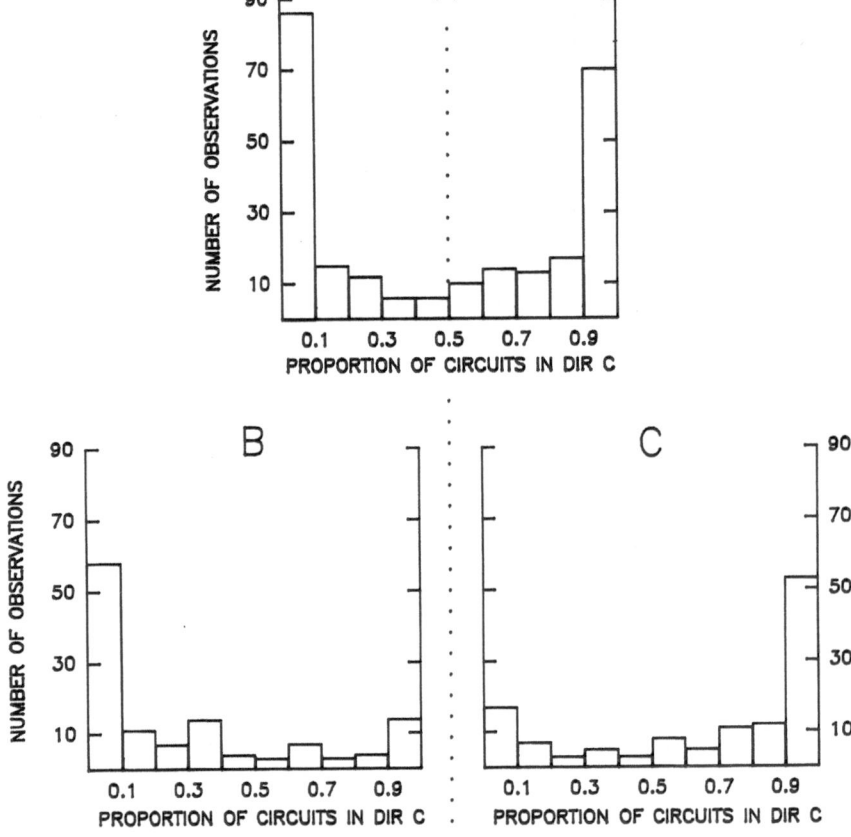

Fig. 2. Frequency distributions of proportions of circuits in Dir C after 100 µmol/kg cocaine. **A** After 1st dose of cocaine. **B** After 2nd dose of cocaine in mice that had a proportion of circuits in Dir C of less than 0.5 after the 1st dose of cocaine. **C** After 2nd dose of cocaine in mice that had a proportion of circuits in Dir C of more than 0.5 after the 1st dose of cocaine

Table 1. Proportions of circuits in Dir C

After 1st dose	After 2nd dose	
	less than 0.5	more than 0.5
Less than 0.5	94	35
More than 0.5	31	89

Chi Sq. = 55, 1 df

total of 1875 circuits were in one direction after the first dose and none in the other direction and then 1924 were in the opposite direction after the 2nd dose and none in the direction of all circuits after the 1st dose. Such a result is astronomically improbable on the basis of chance (Chi square = 40,000; 1 d.f.). High statistical significance would survive any reasonable corrections for post hoc selection. In these mice, asymmetry had no detectable effect. Hence stereotypy alone can determine unidirectional running.

Discussion

As previously reported, 100 µmol/kg cocaine under the circumstances of the present experiments increases the number of circuits made by the mice, and increases the tendency of the mice to run more consistently in either Dir C or in Dir CC.

The tendency to more unidirectional running could be due to individual mice having an asymmetry that biases them to run either in Dir C or in Dir CC, or to cocaine making the mice, having started to run in one direction, simply to continue running in that direction or, as would seem most likely a priori, some quantitative combination of the two factors. The significant correlation between the predominant direction of running for most mice after the two doses of cocaine proves that most mice do have a bias. The bias is not strong in most mice under control conditions (Fig. 1). Even a slight bias, other things being equal, is likely to start a mouse running in the direction of bias; stereotypy could then take over and amplify the consequence of a slight bias.

A substantial minority of mice (66/249) ran strongly in one direction (proportion in Dir C either <0.1 or >0.9) after the 1st dose of cocaine but ran similarly strongly in the other direction after the 2nd dose of cocaine. Clearly, such effects cannot be accounted for by asymmetry; bias if present in these mice must have been overwhelmed by the tendency toward stereotypy. If we assume that the direction of running in these 66 mice was determined largely by chance it would follow that about 66 more mice simply happened to run in the same direction after the 2nd dose of cocaine as they had run after the 1st dose. Thus about 132/249 mice showed no evidence of bias. For the remaining half of the mice, even slight bias would account for- the findings. The present results do not permit quantitative assessment of the relative strengths of the two influences in individual mice and indeed the relative strengths may vary with situation and dose of amphetamine, as indicated by Kokkinidis and Anisman (1977). It is nevertheless clear that stereotypy is a powerful influence even in situations that have been extensively studied as examples of asymmetry.

Acknowledgements

Research supported by USPHS Grants DA 05090 and MH 45641.

References

Cooper SJ, Dourish CT (eds) (1990) Neurobiology of stereotyped behavior. Oxford University Press, Oxford

Dews PB (1990) Directional running in mice: effects of cocaine and chlorpromazine. Psychopharmacology 101: 190–195

Glick SD, Jerussi TP, Zimmerberg B (1977) Behavioral and neuropharmacological correlates of nigrostriatal asymmetry in rats. In: Harnad S, Doty R, Goldstein L, Jaynes J, Krauthamer G (eds) Lateralization in the nervous system. Academic Press, New York, pp 213–249

Glick SD, Hinds PA, Shapiro RM (1983) Cocaine-induced rotation: sex-dependent differences between left- and right-sided rats. Science 221: 775–777

Jerussi TP, Glick SD (1974) Amphetamine-induced rotation in rats without lesions. Neuropharmacology 13: 283–286

Kokkinidis L (1987) Amphetamine-elicited perseverative and rotational behavior: evaluation of directional preference. Pharmacol Biochem Behav 26: 527–532

Kokkinidis L, Anisman H (1977) Perseveration and rotational behavior elicited by d-amphetamine in a y-maze exploratory task: differential effects of intraperitoneal and unilateral intraventricular administration. Psychopharmacology 52: 123–128

Randrup A, Munkvad I, Udsen P (1963) Adrenergic mechanisms and amphetamine induced abnormal behaviour. Acta Pharmacol Toxicol 20: 145–157

Ungerstedt U (1971a) Postsynaptic supersensitivity after 6-hydroxydopamine induced degeneration of the nigro-striatal dopamine system. Acta Physiol Scand [Suppl] 367: 69–93

Ungerstedt U (1971b) Striatal dopamine release after amphetamine or nerve degeneration revealed by rotational behaviour. Acta Physiol Scand [Suppl] 367: 49–68

Author's address: Dr. P. B. Dews, 181 Upland Road, Newtonville, MA 02160, U.S.A.

Subject Index

Supplementum 33

L. Deecke and P. Dal-Bianco (eds.)

Age-associated Neurological Diseases

This volume covers selected up-dated contributions from the Conference on "Age-associated Neurological Diseases" held in Vienna 1989. Fields covered include Age-associated Memory Impairment (AAMI) Alzheimer's and Parkinson's diseases and dementias and movement disorders of other etiologies.

Concerning dementia, some papers deal with diagnosis employing neuro-imaging methods–PET, SPECT, MRI and XCT, others by means of electro-physiological methods. An important aspect is the early preclinical diagnosis of dementia using neuro-psychological tests, to enhance the chance of effective treatment.

Finally, drugs now under clinical investigation are discussed and preliminary results for several compounds are presented. This volume with its up-to-date contributions will be of special interest to all physicians treating elderly persons with Age-associated degenerative diseases.

Springer-Verlag
Wien New York

1991. 30 figures. VIII, 165 pages.
Soft cover DM 98,-, öS 690,-
Reduced price for subscribers to
"Journal of Neural Transmission":
Soft cover DM 89,-, öS 621,-
ISBN 3-211-82261-5

Prices are subject to change without notice.

Supplementum 32

P. Riederer and M. B. H. Youdim (eds.)

Amine Oxidases and Their Impact on Neurobiology

Proceedings of the
4ᵗʰ International Amine Oxidases Workshop,
Würzburg, Federal Republic of Germany, July 7–10, 1990

This book describes the most recent research in the field of monoamine oxidase (MAO), diamine oxidase and semicarbazide-sensitive amine oxidase. MAO and its subtypes have gained enourmous interest in the treatment of a variety of disorders, like depression syndrome Parkinson's disease and possibly dementia of Alzheimer type. Recently selective and reversible MAO-A inhibitors (brofaromine, moclobemide) have been developed. They are safe and easy to handle in clinical practice and show reduced side effects compared to the early irreversible and unspecific MAO-I's. In addition the irreversible and selective MAO-B inhibitor has been shown to prolong life expectancy of patients with Parkinson's disease. Therefore, this book contains additionally to about 36 articles on MAO and its inhibitors, a number of chapters dealing with oxidative stress resulting from radical processes derived in part from deamination. In contrast to other books on the subject, the role of catechol-O-methyltransferase, sulfation and up-take-processes are discussed in detail. This is of particular interest as such an overall view allows a closer insight on neuronal and extraneuronal metabolizing processes.

For the reader the publication is of great interest in so far as the book has been published within four months after hand in of manuscripts. Therefore, the book is a timely written publication of todays research in the field.

Springer-Verlag
Wien New York

1990. 110 figures. XII, 491 pages.
Soft cover DM 240,–, öS 1680,–
Reduced price for subscribers to
"Journal of Neural Transmission":
Soft cover DM 216,–, öS 1512,–
ISBN 3-211-82239-9

Prices are subject to change without notice